EDITOR

Ruth Freitag, Ph.D.
Swiss Federal Institute of Technology,
Lausanne, Switzerland
E-mail:ruth.freitag@epfl.ch
Chapters 3 and 4

CONTRIBUTORS

Miroslav Bleha
Institute of Macromolecular Chemistry
Academy of Science of the Czech Republic
Prague, Czech Republic
E-mail: bleha@imc.cas.cz
Chapter 5

Oliver Brüggemann
Institute for Chemical Engineering
Technische Universität Berlin
Berlin, Germany
E-mail: oliver.brueggemann@berlin.de
Chapter 7

Igor Yu Galaev
Center for Chemistry and Chemical Engineering
Lund University
Lund, Sweden
E-mail: igor.galaev@biotek.lu.se
Chapter 6

Gabriela Grigorescu
Laboratory of Polyelectrolytes and Biomacromolecules
Swiss Federal Institute of Technology
Lausanne, Switzerland
E-mail: gabriela.grigorescu@epfl.ch
Chapter 1

David Hunkeler
Laboratory of Polyelectrolytes and Biomacromolecules
Swiss Federal Institute of Technology
Lausanne, Switzerland
E-mail: david.hunkeler@epfl.ch
Chapter 1

Martin Jordan
Center of Biotechnology
Swiss Federal Institute of Technology
Lausanne, Switzerland
E-mail: martin.jordan@epfl.ch
Chapter 2

Bo Mattiason
Center for Chemistry and Chemical Engineering
Lund University
Lund, Sweden
E-mail: bo.mattiason@biotek.lu.se
Chapter 6

Galya Tishchenko
Institute of Macromolecular Chemistry
Academy of Science of the Czech Republic
E-mail: tishchenko@imc.cas.cz
Chapter 5

Christine Wandrey
Laboratory of Polyelectrolytes and Biomacromolecules
Swiss Federal Institute of Technology
Lausanne, Switzerland
E-mail: christine.wandrey@epfl.ch
Chapter 4

PREFACE

Synthetic polymers fulfill many functions in biotechnology and medicine. In cell culture technology and tissue engineering they provide the surfaces to which cells may attach. Cross-linked polymer networks are used for drug delivery and cell encapsulation. Polymer-based porous membranes can be used to shield implanted cells from the immune system of the host, while allowing for the exchange of nutrients and metabolic waste products thus keeping the cells alive and functioning. In genetic engineering, polymers often play a very important role during the transfer of the foreign genetic material into the recipient cell. In this context polymers present interesting and perhaps safer alternatives to gene delivery by viruses. Last but not least, synthetic polymers have been used to mimic the function of certain biological molecules. Examples are the "artificial antibodies" and "artificial enzymes" produced by a techniques called molecular imprinting. Synthetic displacers in protein displacement chromatography, on the other hand, have to mimic the interaction of the protein with the chromatographic surface to successfully compete for the binding sites and thereby enforce the chromatographic separation.

The idea for this book was first conceived during discussion amongst some of the people at the Swiss Federal Institute of Technology in Lausanne, which use synthetic polymers for some of the above-mentioned purposes. We found that the quality and the properties of these materials were in many cases decisive for the research that could be done with them. For that reason, we thought it might be interesting to outline the needs, the potential and also the state-of-art of some of these domains. While it was sometimes difficult to maintain the enthusiasm, my co-authors and I finally put together this book, which summarizes our knowledge and experience in the use of synthetic polymers in the life sciences. The book starts with two chapters on the delivery of biologicals using synthetic polymers. The chapter on cell encapsulation treats this important subject by taking the bioartificial pancreas as an example. The chapter on gene delivery focuses on the many barriers which nature developed to prevent the genetic modification of cells. Viruses are natural and extremely efficient means of overcome these barriers. Unfortunately, they have in the past given raise to some ethical questions regarding the safety of their use. Artificial polymers will hopefully one day replace these viral systems for the genetic modifications of cells.

The second section of the book deals with the use of synthetic polymers for the purpose of isolating biologicals (bioseparation). The chapter on affinity precipitation describes the use of stimulus-responsive polymers for this purpose. Upon the change of a certain external parameter like the temperature or the pH, such polymers change their behavior, e.g., their solubility in water, in a very abrupt manner. If the polymer is linked to an affinity mediator, any target molecule can be captured and co-precipitated. The issue of stimulus-responsive (sometimes also called "smart") polymers is taken up again in chapter 6. In this chapter a common problem in tissue engineering is addressed. If cells are to be grown on a surface, this surface should have a hydrophilic quality. However, what is good for growth may later become a severe handicap, when the goal is to remove the cells for their final application. Many cells do not react well to the agents commonly used for that purpose. The hydrophobicity of a surface covered with stimulus-responsive polymers, on the other hand, may be changed almost at will by stimulation with a suitable agent. Cells have been known to detach on their own, once a formerly hydrophilic surface had become hydrophobic due to a slight increase in temperature. Other applications of such stimulus-responsive surfaces may be found in bioseparation and drug delivery. The final chapter of the book deals with molecular imprinting as a means to give to polymeric surfaces the ability to distinguish between closely related molecules, which normally is only found in biological compounds such as enzymes.

Certain interesting applications for synthetic polymers in the life sciences are unfortunately not treated in this short book. The use of hybrid molecules (bioconjugates) for drug delivery and other purposes is one example, and the use of polymers in bioseparation by aqueous two-phase systems is another. However, the authors nevertheless hope to have given some indication of the importance of polymeric materials for the life sciences and look forward to future results of the continuous research in this area. As an editor, I would like to thank all contributors to this book for their work and their patience with my sometimes sporadic editing efforts. Last but not least, I would like to thank Ms. Francoise Wyssbrod, who has read and reread (and sometimes retyped) the chapters making sure that they adhered in every detail to the *House Style Manual* provided by the publisher. Without her help, this book would not have been possible.

Ruth Freitag
Lausanne

CHAPTER 1

Cell Encapsulation:

Generalities, Methods, Applications and Bioartificial Pancreas Case Study

Gabriela Grigorescu and David Hunkeler

Introduction

One of the most powerful group of chemicals in the body are organic compounds collectively referred to as hormones. The glands responsible for the production and release of hormones comprise the endocrine system. Endocrine activities have been identified in certain organs, such as the heart, kidneys, duodenum, liver and the islets of Langerhans in the pancreas (which contains the insulin gland), which are normally associated with other system functions.

There have been numerous attempts to replace organ function using cell transplantation including direct injections of dissociated cells into organs such as the liver, kidney or spleen.[1-5] Subcutaneous and intraperitoneal routes have also been evaluated.[6-10] More recent investigations have applied extracellular matrix polymers as structural supports for cell transplantation and immunoprotection.[11,12] Potential medical applications of such "artificial cells" or "tissue engineered" organoids include an extracorporeal bioartificial liver for detoxification,[2] artificial red blood substitutes,[13] the extracorporeal artificial kidney for hemodialysis,[14] immunosorbents[15] and drug delivery systems.[16] The transplantation of immunoisolated (microencapsulated) cells represents another emerging area in biotechnology research and commercialization. Under such a scenario, the encapsulated cells, which could be a xenograft, would be hidden from the immune system of the body, but would still be able to respond to extracellular stimuli (e.g., blood glucose), with the required hormone, in the case of diabetes therapy insulin, secreted into the systemic circulation. Other applications of the microencapsulation concept include the encapsulation of genetically modified cells, which represents a novel approach to somatic gene therapy.[17]

This chapter will review recent advances in cell encapsulation from material science, technological and tissue-related perspectives. Cell coating, microencapsulation devices and bioartificial organs will be discussed with the artificial pancreas and treatment of diabetes used as a case study denominator throughout the review.

Biomaterials

Materials, including synthetic and natural polymers, metals, ceramics and composites have become increasingly important in medicine and pharmaceutics.[18-21] Of these groups,

Synthetic Polymers for Biotechnology and Medicine, edited by Ruth Freitag.

polymers represent the largest class. An extensive classification of the main types of macromolecules according to their origin, properties and fields of application were recently reviewed.[22]

There are three fundamental properties a biomaterial should possess: functionality, mechanical strength, and biocompatibility.[23] The functional characteristic is the specific property required to perform the given task. Mechanical resistance is required to retain an adequate level of device performance, viability and durability in vivo. Finally a "biomaterial" is generally defined as inert material used in a medical device, intended to interact with biological systems[23] which may be used singularly to replace or augment a specific tissue, or in combination to perform a more complex function, e.g., in organ replacement.[24] Biocompatibility is taken to represent the ability of a material to perform with an appropiate host response in a specific application.[25] Biocompatibility can be considered in terms of blood compatibility (hemocompatibility) and tissue compatibility (histocompatibility). Blood compatibility is often defined in terms of events which should not occur, including thrombosis, destruction of formed elements, and complement activation. Histocompatibility encompasses the lack of toxicity and excessive tissue growth around an implant. The biocompatibility of biomedical devices is influenced by the chemical composition of the materials applied, their surface-tissue interactions and by mechanical factors related to the production process.

Most authors[26,27] have described the lack of pericapsular fibrosis (fibroblast overgrowth of the capsule or device) as "biocompatibility". However, local irritation of the environment during the surgical procedure, from the device itself, or an antigen released from the device can induce inflammatory infiltrates which may stimulate the release of substances[26] which are known to be toxic to the tissue to be transplanted. Hence, histological examination of intraperitoneally implanted devices such as microcapsule-based bioartificial organs requires not only removal of the capsules by lavage but also a careful investigation of the peritoneal tissue.

The transplantation of cells for the treatment of variety of human diseases (see Fig. 1.1), such as neurodegenerative disorders or hormone deficiences, has been limited since cells are rapidly destroyed by the recipient's immune system. This is particularly acute for autoimmune diseases such as insulin-dependent diabetes mellitus. Recipient immunosuppression, islet graft pretreatment, and islet transplantation into immunoprivileged sites have not yet provided clinical, or even large animal solutions (islets comprise the endocrine part of the pancreas and contain various cells which produce hormones such as insulin and glucagon in response to chemical stimuli).[10] However, over the past two decades synthetic, semi-synthetic, natural and biological water soluble polymers have been evaluated as potential basic compounds in order to create biomaterials for cell and islet immunoisolation with a variety of materials tolerated intraperitoneally and nontoxic to islets.[28]

Advances in Device and Cellular Engineering

A number of new technologies have been developed and refined during the past several decades which set the stage for a significant advance in transplantation as a major means for treating human disease. These technologies include the identification and isolation of specific cells and cell products which play a major role in disease (hormones, growth factors, immune products, cellular toxins),[30] cell engineering enabling the production of living cells which produce these specific bioactive compounds, and advances in bioreactor design for in vitro maintenance and propagation of these cells.[31] A particular case of encapsulation involves immunoisolation of mammalian cells. Examples include the bioartificial pancreas,[1] enzyme systems[32] and enzyme replacement therapy,[33] encapsulated hepatocytes for the treatment of severe liver failure,[2] the bioartificial kidney,[14] high-density cell growth for immunotherapy,[5] controlled delivery of medicinal substances and other bioactive agents,[34] toxicological studies,[35] entrapment of carcinogens,[36] and hormonal evaluations.[37]

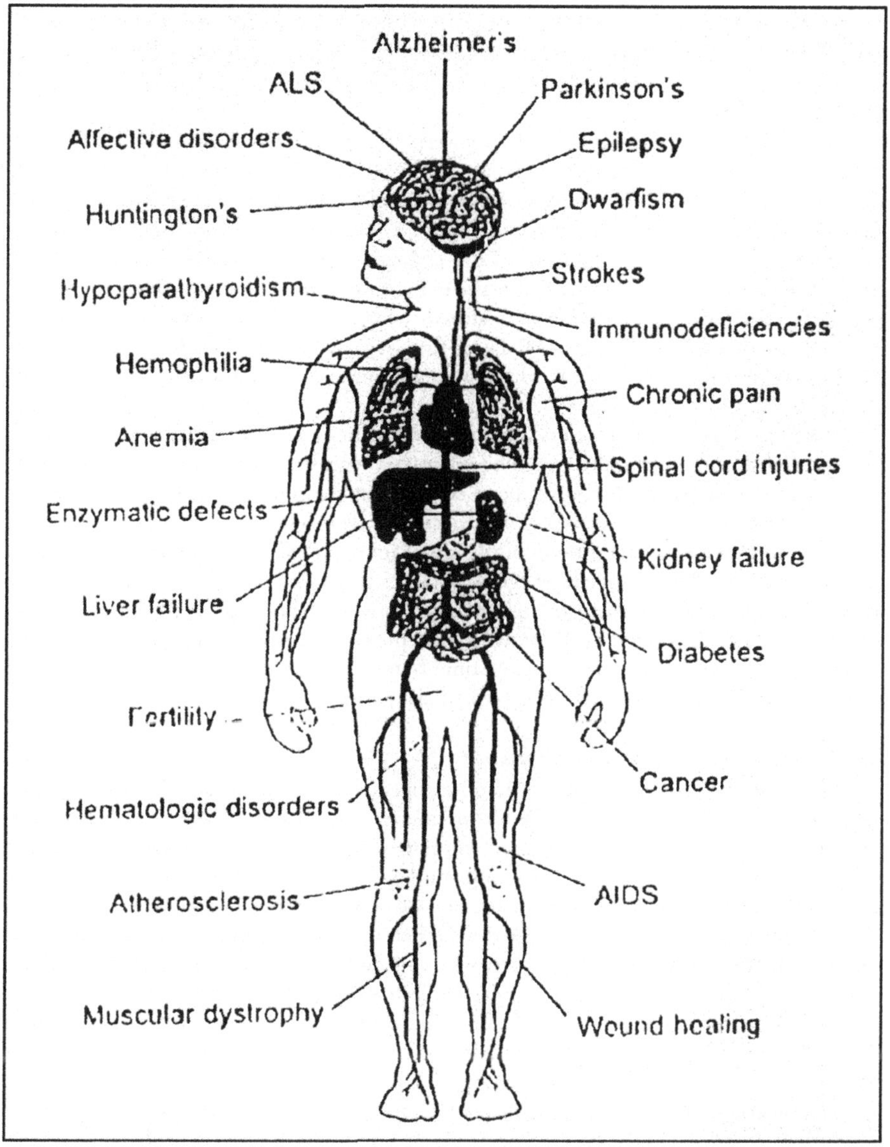

Figure 1.1. Disorders potentially treatable with encapsulated cell transplantation.[29]

The confluence of the aforementioned technologies now enables the development of transplantation beyond whole organs to include specific cells and tissues, which carry out vital differentiated functions. Furthermore, microencapsulation methods have the potential for the treatment of diseases requiring enzyme or endocrine replacement as well as in nutrient delivery of enzymes and bacteria. Encapsulation is also employed in various industries including food,[38,39] agriculture[40,41] and biotechnology.[42] New "intelligent" polymers that respond to small physical or chemical stimuli, such as changes in pH or temperature, glucose[43] or the presence of a specific chemical substrate, have also been synthetised.[44,45]

Immunoisolation

A variety of systems can be employed for cell or enzyme immobilization. These include, for example, microcarriers,[46] gel entrapment,[47] hollow fibers,[48] encapsulation[49] and conformal coatings.[50] The latter three have been extensively tested in small animal models over the last 20 years, particularly in the area of diabetes therapy. The polymeric materials used in bioartificial endocrine devices (the terms bioartificial and endocrine device are often distinguished from 'artificial organs' due to the presence of tissue in the former two) serve two major purposes:

1. as a scaffold and an extracellular matrix they favor the attachment and differentiation of functional cells or cell clusters and keep them separate from one another;
2. as permselective envelopes which provide immunoisolation of the transplant from the host.

The central concept of immunoisolation is the placement of a semipermeable barrier between the host and the transplanted tissue. The properties required for the semipermeable membranes used in cell transplantation depend strongly on the source of cells. An allograft is a transplant between individuals within one species, while a xenograft is a graft between individuals from different species. Immunoisolation of transplanted cells by artificial barriers that permit crossover of low molecular weight substances, such as nutrients, electrolytes, oxygen, and bioactive secretory products, though not of immune cells and high molecular weight proteins such as antibodies (IgG, IgM), provides great promise for developing new technologies to overcome these problems in a reasonable time frame. As an example, Figure 1.2 shows the molecular weight cut off required for a bioartificial pancreas.

Device Geometry Considerations

The immunoisolation of allogeneic or xenogeneic islets can be achieved via two main classes of technology: macroencapsulation and microencapsulation.[51] Macroencapsulation refers to the reliance on larger, prefabricated "envelopes" in which a slurry of islets or cell clusters is slowly introduced and sealed prior to implantation. An intravascular device usually consists of a tube through which blood flows, on the outside of which is the implanted tissue contained within a housing.[52] The device is then implanted as a shunt in the cardiovascular system. Extravascular devices are implanted directly into tissue in a body space such as the peritoneal cavity, though some have also been vascularized into a major artery such as in Calafiore's clinical trial.[53] Geometrical alternatives include cylindrical tubular membranes containing tissue within the lumen and planar diffusion chambers comprised of parallel flat sheet membranes between which the implanted tissue is placed.[54]

Microencapsulation refers to the formation of a spherical gel around each group of islets, cell cluster or tissue fragment. Microcapsules based on natural or synthetic polymers have been used for the encapsulation of both mammalian and microbial cells as well as various bioactive substances such as enzymes, proteins and drugs.[55] A review of alternative semipermeable microcapsules prepared from oppositely charged water soluble polyelectrolyte pairs has been presented in recent papers.[56,57] The main advantage of this approach is that cells, or bioactive agents, are isolated from the body by a microporous semipermeable membrane and the encapsulated material is thus protected against the attack of the immune system. In the case of microencapsulated pancreas islets, a suspension of microcapsules is typically introduced in the peritoneal cavity to deliver insulin to the portal circulation.

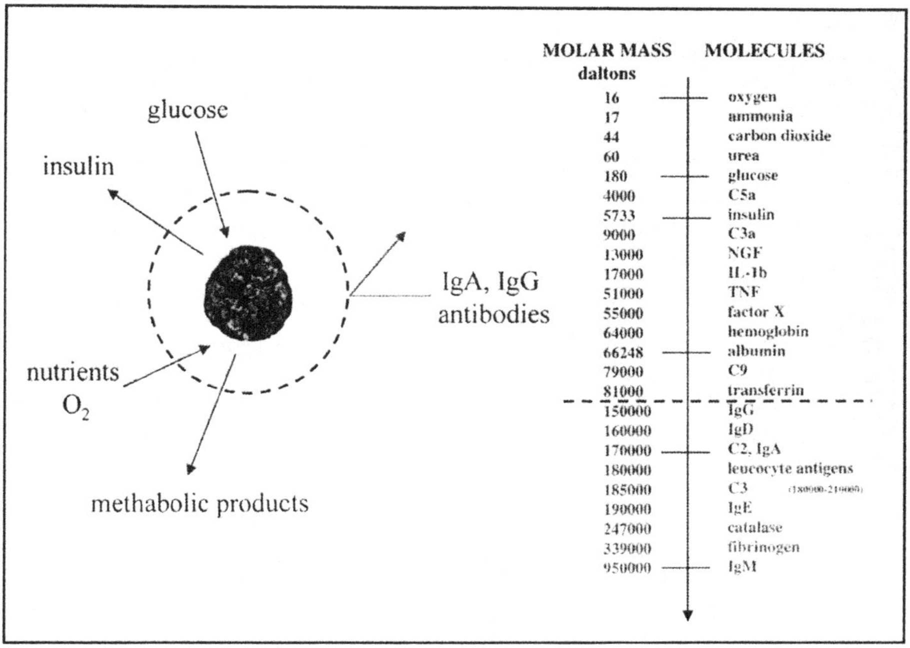

Figure 1.2. Schematic of immunoprotection via a permselective membrane.

Polymer Material Purification, Sterilization and Endotoxin Deactivation

Many commercially available polymers contain impurities which exhibit adverse biological activities and thus may contribute to failure of an allo- or xenograftic implant. These impurities are of several kinds, including monomers, catalysts, and initiators, which are present in synthetically derived polymers. They can usually be removed via dialysis due to their small molecular size. Pyrogens represent the second kind of impurities. They belong to a group of natural compounds of certain gram-negative bacteria (cell wall) and cause fever or sometimes even death when injected intravenously. Chemically, they are represented by a variety of complex lipopolysaccharides with highly hydrophobic character.[57] The third group, mitogens, is a rather less defined class of organic compounds which activate many cell types (including lymphocytes, fibroblasts). Their action leads to cell proliferation and to subsequent production of cytokines involved in inflammatory reactions and implant rejection, if mitogens contaminate polymers used to manufacture such implants.

There are a range of purification methods, including saline precipitation, liquid-liquid separation, two-phase aqueous extraction, polymer precipitation, heat denaturation, isoelectric point separations, dialysis, cheap enough to be use on large volumes of materials. In the case that extreme purity is needed, a further purification[59,60] can be carried using more expensive and complicated methods such as gel filtration, ion exchange, hydrophobic chromatography and displacement chromatography.

Cell Delivery

Each immobilization method has specific properties and advantages. Therefore, the selection of a cell delivery technique depends heavily on the intended application, as will be discussed in the following sections.

Adhesion

Adhesion to a three-dimensional structure is used to immobilize cells for culture or analytical procedures as well as to provide a structural template directing cell growth and differentiation. Adhesion alone does not offer immunoisolation. For in vivo investigations, adhesion-based immobilization must be used in conjunction with either a polymeric membrane or matrix entrapment methods. This method is effective for surface binding, either on top of gel films or within hydrogel foams. Several hydrogels can be engineered with bioadhesive properties by methods which include interfacial polymerization,[61] phase separation,[62] interfacial precipitation[63,64] and polyelectrolyte complexation.[65] Factors affecting cell affinity and behavior on hydrogels include the general chemistry of the monomers and the crosslinks,[66] hydrophilic and hydrophobic properties,[67] and the surface charge and functionality.[68] One method to enhance cell adhesion is by adding immobilized cell-adhesive proteins or oligopeptides, such as the arginine-glycine-aspartic acid sequence, in the hydrogel.[69] The physical characteristics of the hydrogel also govern the adhesion affinity. Therefore, altering the pore size and network structure can modify cell adhesion as well as morphology and function.[70] For some adhesion applications the mechanical strength is also important with a lower fractional porosity generally creating stronger networks. Furthermore, closed pore systems make stronger hydrogels than open pore ones.[71] With the adhesion approach, cells are generally plated onto the hydrogel and allowed to attach and migrate. Supplemented culture media provide the cells with essential nutrients for growth and development as well as a means of oxygen and metabolic product transport while in vitro.

Macroporous hydrogel membranes are manufactured by several techniques. One method of constructing pores large enough for cell growth is by phase separation in the polymer and solvent mixture.[72] The "freeze thaw" and the porosigen techniques are two other approaches.[70] The hydrogel is polymerized around a crystalline matrix made from freezing the aqueous solvent (freeze-thaw technique) or around a porosigen of desired size (porosigen technique). With the "freeze-thaw" method, the ice-based crystalline matrix is then thawed after UV polymerization, leaving a macroporous foam. The porosigen technique also requires removal of the crystalline porosigens, in this case usually by leaching or dispersion after polymerizing of the hydrogel with free-radical initiators has taken place. Another method for constructing hydrogel foams uses gas bubbles from sodium bicarbonate to create the macroporous network.[73] Bubbles are trapped during the gelation stage. Thus, the foam morphology is dependent on the polymerization kinetics and varies for different hydrogel compositions.

Matrix Entrapment

Hydrogels are promising as scaffolds and templates for the entrapment of cells, e.g., for tissue reconstruction and regeneration.[74] Hydrogels are ideal for matrix entrapment since the crosslinks of both synthetic and naturally derived hydrogels provide the essential three-dimensional mesh and porous network to hold the cells in place while allowing the transport of nutrients, wastes and other essential molecules via the bulk fluid. In addition to in vitro applications shared with the adhesion technique, matrix entrapment can be used with in vivo studies to protect transplanted cell-hydrogel complexes from mechanical and immunological damage.[11,12]

Hydrogels for matrix entrapment share some common requirements with polymers chosen for other cell immobilization techniques: biocompatibility to cells and host, selective permeability and good diffusion and transport properties. In addition, hydrogels for matrix entrapment must allow uniform cell distribution. Matrix entrapment hydrogels can be manufactured in various shapes. Gels are often polymerized in situ with the cells in molds or in air or oil (beads).[39] Threads or tube-shaped gels can be manufactured using cylindrical molds. As an extreme example tissue-engineered constructs can be fabricated into the shape of an ear.

Microencapsulation

Microencapsulation is currently the most widely used form of cell delivery with preparation methods including:

1. gelation and polyelectrolyte complexation,
2. interfacial polymerization/phase inversion and
3. conformal coating.

Microencapsulation involves surrounding a collection of cells with a thin generally micrometer sized, semipermeable membrane. Its primary purpose is to protect the encapsulated cells from the host's immune system, while allowing the exchange of small molecules and thereby ensuring cell survival and function. There are several requirements for polymer capsules or hydrogels used as components of microcapsules:

- Noncytotoxicity to the encapsulated cells
- Biocompatibility with the surrounding environment where capsules are to be implanted (e.g., minimal fibrotic response)
- Adequate permeability for diffusion of essential nutrients (e.g., oxygen and glucose for islets of Langerhans) and cell secretory products (such as insulin, metabolic waste)[75]
- Impermeability to secreted antibodies of the host's immune system (e.g., immunoglobulins and glycoproteins after complement activation[75]
- Chemical and mechanical stability

From the technological point of view, the requirements for microencapsulation include:

- Small capsule diameters to ensure sufficient diffusion and internal organ transplantability (depending on application, < 400 μm for bioartificial pancreas),[76] with the cell centering within the microcapsule
- Minimum shrinking/swelling due to changes in osmotic conditions upon transplantation
- Uniform wall thickness for optimum transport of molecules across the membrane and effective immunoprotection.

In addition, the technology used for encapsulation must be nontraumatic to the encapsulated cells. This includes minimizing the mechanical stress during encapsulation and solvent toxicity (if any), as well as optimizing temperature, viscosity, pH and ionic strength. This, in turn, limits the concentration and molecular mass which can be employed. In addition, the ionic content of the polymer backbone (density distribution of charges in the polymer chain), the chemistry and location of functional group attachment, the chain rigidity, aromaticity, conformation and extent of branching were identified as important variables in the type of complex produced. The presence of secondary hydrogen bonding interactions was also found to be significant.

Several problems may prevent wide scale application of microcapsules in the clinic. The capsules can clump together, in which case the cells towards the center may suffer severely from limited diffusion of oxygen and nutrients. A substantial fraction of the capsules may also adhere to tissue. If the capsules degrade, the liberated islet cells, even if nonviable, would greatly increase the antigenic burden on the patient. Semipermeable polymeric membranes have been developed with the aim of permitting the transplantation of xenogenic cells thus removing the need for immunosuppression therapy. However, early clinical implementations is not likely to involve xenografts or genetically modified cells but rather auto- and allografts supplemented by immunosuppression when necessary.

Bioartificial Organs

Tissue engineering involves the in vitro or in vivo generation of organoids such as cartilage, skin or nerves. More ambitious projects seek to ameliorate the quality of life of diseased or injured patients and reduce the economic burden of treatment. Bioartificial organs involve an in vitro prepared tissue-material interface fabricated into a durable device. A typical example is the bioartificial pancreas, which will be discussed in the following section as a case study. The extracorporeal bioartificial liver and more recently the bioartificial kidney[14] are examples of the transient replacement of organ functions, the former intended as a bridge to stabilize comatose patients until a whole organ can be procured. As the bioartificial pancreas is often microcapsule-based, a specific section will be dedicated to review encapsulation technology prior to the application of this bioartificial organ for in situ insulin production.

Bioartificial organs require the combination of several research areas. The understanding of cellular differentiation and growth and how extracellular matrix components affect cell function comes under the umbrella of cell biology. Immunology and molecular genetics will also be needed to contribute to the design of cells or cell transplant systems that are not rejected by the immune system. Cell source and cell preservation are other important issues. The transplanted cells may come from cell lines or primary tissues—from the patients themselves, other human donors, animal sources or fetal tissue. In choosing the cell source, a balance must be struck between ethical issues, safety issues and efficacy. The sterilization and depyrogenation of the polymers involved in transplants is also critical. The materials used in tissue engineering and polymer processing are other key issues. The development of controlled release systems, which deliver molecules over long time periods, will be important in administering numerous tissue-controlling factors, growth factors and angiogenesis stimulators. Finally, it will be useful to develop methods of surface analysis for studying interfaces between cell and materials and mathematical models and in vitro systems that can predict in vivo cellular events.

Microencapsulation Technology

The methods used for microcapsule formation have been recently reviewed.[77] The most widely used procedure involves the gelation of charged polyelectrolytes around the cell core.[78] The popular alginate-L-polylysine microcapsules, for example, are obtained in the following sequence:

1. the cells are embedded in alginate droplets with the aid of a droplet generator (air/liquid jet or an electrostatic generator);[79]
2. the droplets are transformed into rigid beads by inducing cross-linking with calcium ions;
3. the beads are coated with polylysine and alginate, thereby forming the semipermeable capsule; and
4. the alginate core is liquified with a chelating agent.[65]

Microcapsules surrounding individual cells or clusters such as islets should be physically durable, smooth and spherical for optimal biocompatibility. Smoothness is one factor which, in addition to the interfacial composition, reduces tissue irritation, which decreases the probability of cell overgrowth on the capsule surface if aggregated tissue such as beta-cell clusters (beta cells transform blood glucose concentration stimuli into a regulated, pulsatile, insulin secretion) is employed. The capsules should be as small as possible in relation to the islet size to optimize nutrient ingress and hormone egress. Figure 1.3 presents encapsulated rat islets using alginate-cellulose sulphate-poly(methylene-guanidine) microcapsules.

The polyelectrolyte complexation technique used to make alginate-polylysine capsules is advantageous since the capsules are formed under very mild conditions.[78] A disadvantage, however, is the impurities and batch to batch irreproducibility of the alginate, a naturally derived polysaccharide.[80] The high mannuronic acid content of alginate was shown to be responsible for fibrotic tissue response. Fibrosis was reduced and a more resistant microcapsule was fabricated by decreasing the mannuronic acid level of the alginate at the expense of the guluronic acid content,[81] although these conclusions have been questioned by some authors. Another disadvantage of alginate-polylysine microcapsules is that the alginate-polylysine membrane, a weak polyelectrolyte complex, gives the microcapsules relatively poor mechanical properties.

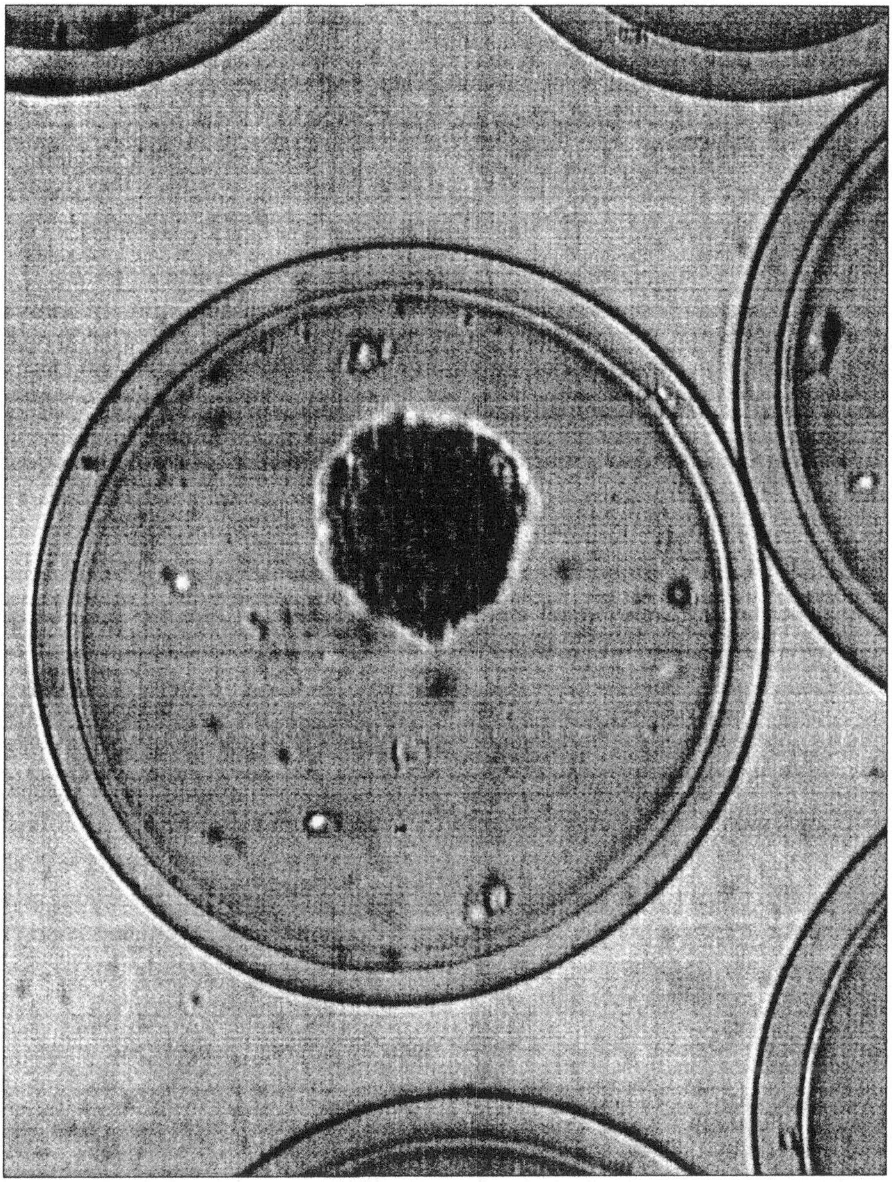

Figure 1.3. Alginate-cellulose sulphate-poly(methylene-guanidine) microcapsules containing rat islets.

Local changes in pH or ionic concentration may have influence on the integrity of these microcapsules drastically.[78]

Several different hydrogels have been investigated to determine the efficacy of encapsulation therapy as treatment for multiple diseases in a variety of animal models. For instance, alginate-polylysine-alginate microcapsules have been employed to encapsulate islets and to reverse the effects of diabetes in rats and mice.[82] The mild encapsulation procedure preserved the integrity of the islet's secretory function with long-term viability maintained.[83] Modified alginate-polylysine microcapsules, which are smaller and stronger than the previous versions, improved the survival of the xenographic tissue grafts.[78] Coating alginate-polylysine capsules with a poly(ethylene glycol)hydrogel[84] or incorporating monomethoxy poly(ethylene glycol) pendant chains to the polylysine polymer backbone[85] has led to improved biocompatibility compared to unmodified capsules. In an attempt to simultaneously control biocompatibility and permeability, polymer blends have been selected that were optimal with respect to islet cytotoxicity (as measured by in vivo tests or) as well as thermodynamic (swelling/shrinking) and mechanical parameters.[86,87]

Interfacial polymerization is another method developed for encapsulation of mammalian cells. Cells are coextruded with a generally hydrophobic polymer solution through a coaxial needle assembly. Shear and mechanical forces due to a coaxial air/liquid stream flowing past the tip of the needle assembly causes the hydrogel to envelop the cells and fall off. The encapsulated cells fall subsequently through a series of oil phases, which cause precipitation of the hydrogel around the cell. This process, based on membrane phase inversion, is used primarily when encapsulating cells with hydrogels from the polyacrylate family.[88] Polyacrylates are well tolerated by the host's immune system and have exceptional hydrolytic stability.[88] A potential disadvantage of this technique is that organic solvents, which may be harmful to living cells, are used to precipitate the hydrogel. To eliminate the use of organic solvents, complex coacervation was developed using acidic and basic water-soluble polymers.[88] Briefly, a droplet containing one of these polymers and cells is added to the other polymer. A thin membrane encapsulates the droplet due to ionic interactions of the two polymers. The major disadvantage of this method is that the capsules may be unstable due to high water uptake in the capsule wall. Modifications have been made to better control permeability and stability of the hydrogel capsules.[88]

Photopolymerization has also been used to conformally coat hydrogel capsules to:

1. improve their biocompatibility and
2. reduce the volume to a minimum in order to reduce implant size, a critical issue if an internal organ is the intended transplantation site.[89]

Photopolymerization permits gelation of the polymer membrane in the presence of dissolved oxygen, which is helpful for cell survival during the encapsulation process. The advantage of this technique is that the membrane is directly in contact with the encapsulated cells. Minimizing diffusion distance for oxygen, nutrients, and cell products is important for eliminating necrosis at the center of the capsule[12] and for improving therapeutic efficiency.

Case Study: Insulin Production Systems

Type I diabetes mellitus is a disorder affecting over 80 million people worldwide. At present exogenous insulin delivery via injection or pumps equipped with glucose sensors cannot provide the minute-to-minute normoglycemia needed to prevent the complication associated with this autoimmune disorder. The sensor pump technology also lacks durability, with device function often limited to only hours. The exacting requirement placed on insulin dosage and timing of administration in diabetic patients, as well as the many years of safe and reliable treatments expected from the insulin delivery technology, have pointed to the advantages of implantable systems in which insulin would be synthesized as needed and made available to the organism on demand. Four alternatives have been considered and have undergone clinical evaluation: whole organ transplantation, human islet and xenogeneic islet transplantation, immunoisolation of normal or tumoral insulin-secreting tissue, and transplantation of genetically-engineered cells to replace the functions of the beta cells.

At present there are three critical problem areas in the further development of implantable immunoisolation devices:

1. supply of tissue,
2. device design and performance, and
3. protection from immune rejection.

These will be discussed in the following sections.

Tissue Sourcing

Organs and cells of animal origin are being considered as a source of tissue for xenotransplantation.[90,91] If islet transplantation is to become a widespread treatment for type I diabetics, solutions must be found for increasing the availability of insulin-producing tissue and for overcoming the need for continuous immunosuppression. Insulin-producing cells being considered for clinical transplantation include porcine and bovine islets, fish-Brockman bodies,[92] genetically engineered insulin-secreting cell lines and in vitro produced "human" β-cells.

Both primary tissue and cultured cell lines have been employed in small animal xenotransplantation, including cells that have been genetically modified.[93] Substantial efforts have also been made in the isolation of primary tissue, especially for pancreatic islets,[94] though further improvements are necessary for practical, large-scale processing. The most urgent problem in transplantation is the shortage of donor organs and tissue. Xenotransplantation could offer some advantages over the use of human organs. Xenotransplantation could be planned in advance, the organ would be transplanted while it was still fresh and undamaged. In addition, a planned transplantation allows the administration of therapeutic regimens that call for the pretreatment of the recipient. Another advantage is the possibility that animal sources could be genetically engineered in order to lower the risk of rejection by expressing specific genes for the benefit of the patient. However, the concern over retroviruses has led to political moratoriums on the clinical use of xenotransplantation. It has yet to be established in nonrodent models as a viable alternative.

Cell Banking and Transplanted Tissue Volume

Certain human cells[95-98] can be readily cultivated and scaled up for cell banking (cells are taken from an animal and cultured in vitro under specific conditions to greatly expand the amount of tissue available). A partial list includes: skin cells, vascular cells, adipose tissue cells, skeletal muscle cells, chondrocytes, osteoblasts, mucogingeval cells, corneal cells, skeletal muscle cells and pigment cells. Roughly 450,000 human islets, or about 6,500 islets per kilogram body weight, should be adequate to provide normal blood glucose control.[2] However, islet requirements in published studies have ranged from a low of about 3,500 islets/kg to as much as 30,000 to 60,000 islets/kg. These large values in some studies suggest that many of the islets in some implanted immunoisolated devices are either not viable or not functioning at their normal level.

Alternative Tissue Sources

The optimal source of xenogeneic islets remains controversial. Islets have been isolated from primates and xenografted into immunosuppressed, diabetic rodents, with short-term reversal of diabetes.[98] However, there are ethical issues surrounding the use of primates for these studies. Other promising islet sources are porcine, bovine and rabbit islets, all of which function remarkably well in diabetic rodents.[99] Long-term human, bovine and porcine islet xenograft survival has been documented in nude mice and rats, suggesting that, in the absence of an immune response, sufficient islet-specific growth factors are present in xenogeneic recipients.[100]

Porcine islets are at present receiving the greatest attention since pigs produce an insulin which is structurally very similar to human insulin and pigs are, on the other hand, the only large animals slaughtered in sufficient quantities to supply the estimated demand from type I diabetics.[101,102] In addition, porcine islets within microcapsules have been reported to correct diabetes in cynomologus monkeys.[103] Elaborate studies are in progress to engineer a "perfect pig", having adequate levels of complement-inhibiting factors.[104] Thus, porcine sources are perhaps most likely to provide islets for an inaugural human xeno-islet trial. However, porcine islets are fragile and have poor long-term stability. The in vitro glucose-stimulated insulin secretion rate per unit islet volume appears to be substantially smaller for porcine islets than for other species including human. Lastly, there is significant current concern regarding the potential for transmission of infectious agents from porcine organ sources to human xenograft recipients, and to the population at large.[105,106] None of these characteristics bode well for their practical large-scale use, and serious consideration and investigation is being given to alternate animal sources. There is also speculation that neonatal porcine islets, which culture better and present minimal infrastructure problems, would be an ultimate substitute.[107] Isolation of bovine islets is technically easier and calf islets are glucose-responsive.[101] However, adult bovine islets are relatively insensitive to glucose.[101] The rabbit pancreas is also an attractive source of islets since rabbit insulin differs from human insulin at only one amino acid and rabbit islets are glucose responsive.[99]

Another approach of recent interest is development of a so-called artificial β cell by use of recombinant DNA techniques. Such a genetically engineered cell line must sense glucose concentration and secrete insulin appropriately at a rate per unit islet volume that is comparable to primary tissue.

Islet Viability and Function

The permeability of immunoisolation devices must balance two potentially conflicting requirements. First, cells enclosed within the device must receive all the molecules and factors necessary for viability and normal function. Secondly, the destructive components of the immune system should be prevented from entering the immunoisolation device. Lymphocytes and macrophages are easily excluded by all immunoisolation devices; however, many soluble products of the immune system such as complement protein, cytokines and nitric oxide may also be cytotoxic to immunoisolated cells. Islets of Langerhans in vivo are highly vascularized by a network of capillaries that deliver nutrients and oxygen to each beta cell. However, in the immunoisolation state, vascular assess to the islet is eliminated, and solutes move to and from the islet cells by diffusion from the surrounding environment. The diffusion gradients of wastes, nutrients, and especially oxygen are important.

The oxygen levels to which the islet cells are exposed are important from two standpoints, viability and function. Because oxygen is consumed at a high rate by islet cells, particularly when stimulated by increased glucose concentration, steep gradients in oxygen concentration can develop. Thus, the oxygen concentration decreases from that of the local blood supply as it diffuses across the tissue, the immunoisolation membrane, and throughout the islet. Consequently, islet cells may be exposed to hypoxic, or even anoxic, conditions.[60] This can lead to loss of cell viability and to a reduction in the insulin secretion capacity.[108] Further studies should focus on finding a practically applicable method to reduce the barrier between encapsulated islets and the bloodstream in order to improve both the functional performance and the survival of encapsulated islet grafts. However, an interchange between vascularization and hence nutrient supply and retrievability will always be present.

Bioartificial Organ Rejection

The process of rejection may begin with the diffusion of immunogens from the graft across the membrane barrier. There are several possible sources for these antigens, including molecules shed from the cell surface, protein secreted by live cells and cytoplasmic protein liberated from dead cells. Recognition and display of these antigens by antigen presenting cells initiates the cellular and humoral immune response. The former leads to activation of cytotoxic cells, macrophages and other cells of the immune system. These cells must be prevented from contacting grafted tissue, a requirement relatively easy to meet. More difficult is keeping out components of the humoral immune response. These include cytokines, for example, interleukin-1, which can have detrimental effects on beta cells, as well as the antibodies formed as a response to the antigens, which have leaked across the barrier. In addition, there may always be some antibodies already present in the antibody spectrum of the blood serum which correspond to cell surface antigens (e.g., major histocompatibility complexes) on allo- or xenografts. Antibodies produced during preexisting autoimmune disease, such as type I diabetes, might also bind to surface antigens on allogeneic cells. Finally, macrophages and certain other immune cells can secrete low-molecular weight reactive metabolites of oxygen and nitrogen including free radicals, hydrogen peroxide, and nitric oxide that are toxic to cells in a nonspecific manner. These agents can diffuse large distances if their lifetime exceeds 1 s.[6]

Any attempts to evaluate biocompatibility in vitro would show some lack of predictability for in vivo experiments. Therefore, implantation experiments are necessary to correlate these phenomena. The majority of experiments have been performed on rodents,[26] and there are only a few reports on systematic experiments in large animal models.[109] The choice of an animal model should reflect the human situation. In diabetes research, the diabetic BB-rat, NOD-mice and STZ-treated mice have generally been accepted to be a representative animal model of autoimmune diabetes.[27,110]

Implantation Sites

None of the currently reported sites employed for islet transplantation, i.e., the liver,[6,7] the spleen,[7,8] beneath the renal capsule[7,9] and the omental porch[111] and the peritoneal cavity,[10] combine the capacity to bear high numbers of islets and retrievability of the islet graft. However, a site with both features may be a mandatory for large-scale clinical transplantation of encapsulated islets, because such grafts still have their functional limitations and, therefore, may require repeated replacement. Recently, the concept of an intraperitoneally implanted solid support as a transplantation site for genetically-engineered cells has been proposed.[112] Such a solid support may serve as a transplantation site for pancreatic islets as, theoretically, it allows for implantation of high numbers of islets that can be readily retrieved.

Socio-Political Considerations

The application of microencapsulated cells provides a flexible therapy for transplantation, subcutaneous insertion, extracorporeal perfusion and oral administration. However, organ transplantation evokes ethical questions. Scarcity of donor organs implies that the waiting lists of potential recipients for certain organs is growing. This is particularly true for the kidney. The number of patients dying while on the waiting list also increases with time. Moreover, among the potential donors the number of cadaveric organs utilized is further reduced following complications of sustained intensive care. The issue of multiple transplantation for a single recipient at the expense of those of the waiting list is also an issue.

The need for an alternative source of organs, together with the expansion of scientific data in this field, has focused attention on xenotransplantion as a possible alternative to allotransplantation in the treatment of patients with end-stage disease of vital organs. The spread of animal-derived pathogens to the recipient and to the general population, termed "xenosis", is a potential complication of interspecies transplantation.[105,106] Regulatory and public health agencies, as well as scientific and medical organizations, have held numerous meetings addressing this issue. The UK, Switzerland and the USA have recently placed limited moratoriums on xenotransplantation.[105,106] Fetal tissue sources are under consideration, though these present ethical challenges, particularly with respect to human tissue.

The reproducible isolation and preservation of functional islets on a large scale remains difficult, costly and laborious. Cells used in a bioartificial organ may be stored (e.g., cryopreserved)[113] and screened for adventitious agents prior to use. Tissue storage and the use of a selective membrane are two key differences between bioartificial organs and xenotransplantation and may help reduce the risk of zoonosis.[114] To deal with supply-related issues, centers of excellence in cryosuppression have been proposed. However, it remains to be determined if and how banking will be coordinated on a municipal, regional, national or continental scale.

Conclusions

Current methods of transplantation and tissue reconstruction are among the most costly clinical therapies. Furthermore, the treatment of the secondary effects of diseases such as diabetes contributes significantly to the annual public expenditure in developed and emerging regions. Cell delivery offers the possibility of substantial future savings by providing substitutes that are less expensive than donor organs and the excessive medical following required. In addition, cell transplant systems may complement gene therapy approaches in facilitating transfer of large populations of cells expressing a desired phenotype. Research oriented at novel materials development, in vitro organoid synthesis as well as large scale tissue sources via discordant xenografts and genetically-engineered cells remain promising areas for public and private investment. Socio-politically both are likely to be preceded by demonstration technologies based on allografts, which target the worst case 10-20% of patients.

References

1. Sun AM. Microencapsulation of pancreatic islets—A bioartificial endocrine pancreas. Methods Enzymol 1988; 137:575-580.
2. Iwata H, Dajiki T, Maeda H et al. In vitro evaluation of metabolic functions of a bioartificial liver. ASAIO 1999; 45:299-306.
3. Koebe HG, Muhling B, Deglmann CJ et al. Cryopreserved porcine hepatocyte cultures. Chem Biol Interact 1999; 121:99-115.
4. Zhou D, Kintsourashvili E, Mamujee S et al. Bioartificial pancreas: Alternative supply of insulin-secreting cells. Ann NY Acad Sci 1999; 875:208-218.
5. Reilly EB, Antognetti G, Wesolowski JS et al. The use of microcapsules for high density growth of human infiltrating lymphocytes and other immune reactive T cells. J Immunol Methods 1990; 126:273-279.
6. Lorenz D, Rosenbaum KD, Petermann J et al. Transplantation of isologous islets of Langerhans in diabetic rats. Acta Diabetol Lat 1975; 12:30-40.
7. Van Suylichem PT, Strubbe JH, Houwing H et al. Insulin secretion by rat islet isografts of a defined endocrine volume after transplantation to three different sites. Diabetologia 1992; 35:917-923.
8. Feldman SD, Hirshberg GE, Dodi G et al. Intrasplenic islet isografts. Surgery 1977; 82:386-394.
9. Lacy PE, Davie JM, Finke EH. Transplantation of insulin-producing tissue. Am J Med 1981; 70:589-594.
10. Fritschy WM, van Straaten JF, De Vos P, et al. The efficacy of intraperitoneal pancreatic islet isografts in the reversal of diabetes in rats. Transplantation 1991; 52:777-783.
11. Lanza RP, Sullivan SJ, Chick WL. Islet transplantation with immunoisolation. Diabetes 1992; 41:1503-1510.
12. Mikos AG, Papadaki MG, Kouvroukoglou S et al. Islet transplantation to create a bioartificial pancreas. Biotechnol. Bioeng 1994; 43:673-677.
13. Chang TM. Artificial cells, encapsulation, and immobilization. Ann NY Acad Sci 1999; 875:71-83.
14. Humes HD, MacKay SM, Funke AJ et al. Tissue engineering of a bioartificial renal tubule assist device: In vitro transport and metabolic characteristics. Kidney Int 1999; 55:2502-2514.
15. Benny WB, Sutton DM, Oger J et al. Clinical evaluation of a staphylococcal protein A immunoadsorption system in the treatment of myasthenia gravis patients. Transfusion 1999; 39:682-687.

16. Shaw RJ. Inhaled corticosteroids for adult asthma: Impact of formulation and delivery device on relative pharmacokinetics, efficacy and safety. Respir Med 1999; 93:149-160.
17. Cioffi L, Sturtz F, Wittmer S et al. A novel endothelial cell-based gene therapy platform for the in vivo delivery of apolipoprotein E. Gene Ther 1999; 6:1153-1159.
18. Wang M, Hench LL, Bonfield W et al. Bioglass/high density polyethylene composite for soft tissue applications: preparation and evaluation. J Biomed Mater Res 1998; 42:577-586.
19. Tan J, Saltzman WM. Influence of synthetic polymers on neutrophil migration in three-dimensional collagen gels. J Biomed Mater Res 1999; 46:465-474.
20. Wang M, Joseph R, Bonfield W. Hydroxyapatite-polyethylene composites for bone substitution: Effects of ceramic particle size and morphology. Biomaterials 1998;19:2357-2366.
21. Taylor M, Tanner KE, Freeman MA. Finite element analysis of the implanted proximal tibia: A relationship between the initial cancellous bone stresses and implant migration. J Biomech 1998; 31:303-310.
22. Angelova N, Hunkeler D. Rationalizing the design of polymeric biomaterials. TRIB 1999; 17:27-38.
23. Courtney JM, Lamba NMK, Gaylor JDS et al. Blood-containing biomaterials: Bioengineering viewpoints. Artif Organs 1995; 19:852-856.
24. Williams DF, ed. Definitions in Biomaterials; Elsevier Science Ltd., 1987.
25. Friedman DW, Orland PJ, Greco RS. History of biomaterials. In: Greco RS, ed. Implantation Biology—The Host Responses and Biomedical Devices. Boca Raton: CRC Press, 1994:1-65.
26. Cole DR, Waterfall M, McIntre M et al. Microencapsulated islet grafts in the BB/E rat: A possible role for cytokines in graft failure. Diabetologia 1992; 35:231-237.
27. Wijsman J, Atkinson P, Mazaheri R et al. Histological and immunopathological analysis of recovered encapsulated allogeneic islets from transplanted diabetic BB/W rats. Transplantation 1992; 54:588-592.
28. Lamba NMK, Woodhouse KA, Cooper SL. Polyurethanes. In: Biomedical Applications. Boca Raton: CRC Press, 1998.
29. Lanza RP, Hayes JL, Chick WL. Nature Biotechnol 1996; 14:1107-1111.
30. Reiner G, Hecht W, Leeb T et al. Isolation and characterization of the porcine c-myc proto-oncogene and chromosomal assignment to SSC 4p13. Anim Genet 1999; 30:204-206.
31. Kue R, Sohrabi A, Nagle D et al. Enhanced proliferation and osteocalcin production by human osteoblast-like MG63 cells on silicon nitride ceramic discs. Biomaterials 1999; 20:1195-1201.
32. Parthasarathy RV, Martin CR. Synthesis of polymeric microcapsule arrays and their use for enzyme immobilization. Nature 1994; 369:298-301.
33. Rudolph AS, Sulpizio A, Hieble P et al. Liposome encapsulation attenuates hemoglobin induced vasoconstriction in rabbit arterial segments. J Appl Physiol 1997; 82(6):1826-1835.
34. Braybrook JH, Hall LD. Polymeric controlled release systems. Drug Des Deliv 1990; 6:73-86.
35. Collins TF, Sprando RL, Shackelford ME et al. Food and drug administration proposed testing guidelines for developmental toxicity studies. Regul Toxicol Pharmacol 1999; 30:39-44.
36. Buboltz JT, Feigenson GW. A novel strategy for the preparation of liposomes: Rapid solvent exchange. Biochim Biophys Acta 1999; 1417:232-245.
37. Hunt SM, Clarke CL. Expression and hormonal regulation of the *sox4* gene in mouse female reproductive tissues. Biol Reprod 1999; 61:476-481.
38. Perols C, Piffaut B, Scher J et al. The potential of enzyme entrapment in konjac cold-melting gel beads. Enzyme Microbiol Technol 1997; 20(1):57-60.
39. Groboillot A, Boadi DK, Poncelet D et al. Immobilization of Cells for application in the food industry. Crit Rev Biotech 1994; 14(2):75-107.
40. Chung YC, Huang CP, Tseng CP. Biotreatment of ammonia from air by an immobilized arthobacter oxydans CH8 biofilter. Biotechnol Prog 1997; 13(6):794-798.
41. Gimeno M. An overview of the latest development of microencapsulation for agricultural products. J Environ Sci Health 1996; B31(3):407-420.
42. Chang TMS. Artificial cells with emphasis on bioencapsulation in biotechnology. Biotechnol Ann Rev 1995; 1:267-295.
43. Kagatani S, Shinoda T, Konno Y et al. Electroresponsive pulsatile depot delivery of insulin from poly(dimethylaminopropylacrylamide) gel in rats. J Pharm Sci 1997; 86:1273-1277.

44. Bromberg LE, Ron ES. Temperature-responsive gels and thermogelling polymer matrices for protein and peptide delivery. Adv Drug Deliv Rev 1998; 31:197-221.
45. Shiin D, Murata Y, Kataoka K et al. Preparation and characterization of a glucose-responsive insulin-releasing polymer device. Biomaterials 1994; 15:121-128.
46. Shahar A, Reuveny S. Nerve and muscle cells on microcarriers in culture. Advances in Biochemical Engineering/Biotechnology. Vol. 34. 1987; 34:33-35.
47. Guo Y, Lou F, Peng Z et al. Kinetics of growth and α-amylase production of immobilized *Bacillus subtilis* in an airlift bioreactor. Biotech Bioeng 1990; 35:99-102.
48. Shukla R, Kang W, Sirkar KK. Acetone-butanol-ethanol production in a novel hollow fiber fermentor-extractor. Biotech Bioeng 1989; 34:1158-1166.
49. Lim F, Encapsulating Biological Materials. US Patent, 1982; No. 4,352,883.
50. May MH, Sefton MV. Conformal coating of small particles and cell aggregates at a liquid-liquid interface. Ann NY Acad Sci 1999; 875:126-34.
51. van Schilfgaarde R, de Vos P. Factors influencing the properties and performance of microcapsules for immunoprotection of pancreatic islets. J Mol Med 1999; 77:199-205.
52. Petruzzo P, Pibiri L, De Giudici MA et al. Xenotransplantation of microencapsulated pancreatic islets contained in a vascular prosthesis: Preliminary results. Transpl Int 1991; 4:200-204.
53. Kallmes DF, Williams AD, Cloft HJ et al. Platinum coil-mediated implantation of growth factor-secreting endovascular tissue grafts: An in vivo study. Radiology 1998; 207:519-523.
54. Maki T, Mullon CJ, Solomon BA et al. Novel delivery of pancreatic islet cells to treat insulin-dependent diabetes mellitus. Clin Pharmacokine. 1995; 28:471-482.
55. Goosen MFA. Fundamentals of animal cell encapsulation and immobilization. Boca Raton: CRC Press, 1993.
56. Prokop A, Hunkeler D, DiMari S et al. Water soluble polymers for immunoisolation I. Complex Coacervation and cytotoxicity. Adv Polym Sci 1998; 136:1-51.
57. Prokop A, Hunkeler D, Powers AC et al. Water soluble polymers for immunoisolation II: Evaluation of multicomponent microencapsulation systems. Adv Polym Sci 1998; 136:53-73.
58. Nowotny A. Review of the molecular requirements of endotoxin actions. Revs Infect Dis 1987; 9(5):503-511.
59. Prokop A, Wang TG. Purification of polymers used for fabrication of an immunoisolation barrier. In: Prokop A, Hunkeler D, Cherrington AD, eds. Bioartificial Organs. New York: NY Acad Sci, 1997:223-231.
60. Zimmerman U, Klock G, Federlin K et al. Horcher A, Entemann H, eds. Production of mitogen contamination free alginate with variable ratios of mannuronic acid to guluronic acid by free flow electrophoresis. Electrophoresis 1992; 13:269-274.
61. Chang TMS. Semipermeable microcapsules. Science 1964; 146:524-525.
62. Green BK. Pressure-Sensitive Record Materials. US Patent No. 2, 1955; 712, 507.
63. Sefton MV, Crooks CA, Horvath V et al. HEMA-MMA copolymer for the microencapsulation of mammalian cells. Research Report 1990, University of Toronto.
64. Park K, Shalaby WS, Park H. Biodegradable hydrogels for drug delivery. Lancaster: Technomic, 1993.
65. Sun AM, Cai Z, Shi Z et al. Microencapsulated hepatocytes-an in vitro and in vivo study. Biomater Artif Organs 1987; 15:483-496.
66. Lanza RP, Jackson R, Sullivan A et al. Xenotransplantation of cells using biodegradable microcapsules. Transplantation 1999; 67(8):1105-1111.
67. Smetana K Jr. Cell biology of hydrogels. Biomaterials 1993; 14:1046-1050.
68. Ratner B. Hydrogel surfaces. In: Peppas NA, ed. Hydrogels in Medicine and Pharmacy: Fundamentals. Boca Raton: CRC Press, 1986:85-94.
69. Drumheller PD, Hubbell JA. Surface immobilization of adhesion ligands for investigations of cell/substrate interactions. In: Bronzino JD, ed. Handbook of Biomedical Engineering. Boca Raton: CRC, 1995:1583-1592.
70. Oxley HR, Corkhill PH, Fitton JH et al. Macroporous hydrogels for biomedical applications: Methodology and morphology. Biomaterials 1993; 14:1064-1072.
71. Gibson L, Ashby M. Cellular Solids: Structure and Properties. New York: Pergamon Press, 1988.

72. Woerly S. Hydrogels for neural tissue reconstruction and transplantation. Biomaterials 1993; 14:1056-1058.
73. Chen J, Park K. Synthesis and characterization of superporous hydrogel composites. J Controlled Release 2000; 65:73-82.
74. Woerly S, Marchand R, Lavallee C. Intracerebral implantation of synthetic polymer/biopolymer matrix: A new perspective for brain repair. Biomaterials 1990; 11:97-107.
75. Morris PJ. Immunoprotection of therapeutic cell transplants by encapsulation. Trends Biotechnol 1996; 14:163-167.
76. Basta G, Rossodivita ME, Osticioli L. et al. Ultrastructural examination of pancreatic islet containing alginate/polyaminoacidic coherent microcapsules. J Submicr Cytol Pathol 1996; 28:209-213.
77. Renken A, Hunkeler D. Microencapsulation: A review of polymers and technologies with a focus on bioartificial organs. Polymery 1998; 43:530-539.
78. Aebischer P, Goddard M, Trasco P. Cell encapsulation for the nervous system. In: Goosen MFA, ed. Fundamentals of Animal Cell Encapsulation and Immobilization. Boca Raton: CRC Press 1993:197-224.
79. Lum Z, Tai I, Krestnow M. et al. Prolonged reversal of diabetic state in NOD mice by xenografts of microencapsulated rat islets. Diabetes 1991; 40:1511-1516.
80. Sun AM, Vacek I, Tai I. Microencapsulation of living cells and tissues. In: Dunbrow M, ed. Microcapsules and Nanoparticles in Medicine and Pharmacy. Boca Raton: CRC Press 1992:315-322.
81. Soon-Shiong P, Feldman E, Nelson R et al. Long-term reversal of diabetes by the injection of immunoprotected islets. Proc Nat Acad Sci 1993; 90:5843-5847.
82. Fan M, Lum Z, Fu X. et al. Reversal of diabetes in BB rats by transplantation of encapsulated pancreatic islets. Diabetes 1990; 39:519-522.
83. Levesque L, Brubaker PL, Sun AM. Maintenance of long-term secretory function by microencapsulated islets of Langerhans. Endocrinology 1992; 130:644-650.
84. Sawhney AS, Pathak CP, Hubbell JA. Interfacial photopolymerization of poly(ethylene glycol)-based hydrogels upon alginate-poly(L-lysine) microcapsules for enhanced biocompatibility. Biomaterials 1993; 14:1008-1017.
85. Sawhney AS, Hubbell JA. Poly(ethylene oxide)-graft-poly(L-lysine)-alginate microcapsules membranes. Biomaterials 1992; 13:863-870.
86. Prokop A, Hunkeler D, DiMari S. Adv Polym Sci 1998; 136:1-51.
87. Wang T, Lacik I, Brissova M et al. An encapsulation system for the immunoisolation of pancreatic islets. Nature Biotechnol 1997; 15:358-362.
88. Sefton MV, Stevenson WTK. Microencapsulation of live animal cells using polyacrylates. Adv Polym Sci 1993; 107:143-197.
89. Pathak CP, Sawhney AS, Hubbell JA. Rapid photopolymerization of immunoprotective gels in contact with cells and tissue. J Am Chem Soc 1992; 114:8311-8312.
90. Edge AS, Gosse ME, Dinsmore J. Xenogeneic cell therapy: Current progress and future development in porcine cell transplantation. Cell Transplant 1998; 7(6):525-539.
91. Lin S, Saadi S. New development in solid organ xenotransplantation. Graft 1998; 1(2):45-47.
92. Wright JR, Polvi S, Maclean H. Experimental transplantation with principal islets of teleost fish (Brockman Bodies). Long-term function of tilapia islet tissue in diabetic nude mice. Diabetes 1992; 41:1528-1532.
93. Kume A, Ito K, Ueda Y et al. A G-CSF receptor-gyrase B fusion gene: A new type of molecular switch for expansion of genetically modified hematopoietic cells. Biochem Biophys Res Commun 1999; 260:9-12.
94. Lanza RP, Chick WL. Pancreatic islet transplantation. Vol. I. Procurement of pancreatic islets. Austin: R.G. Landes, 1994.
95. Richard L, Velasco P, Detmar M. A simple immunomagnetic protocol for the selective isolation and long-term culture of human dermal microvascular endothelial cells. Exp Cell Res 1998; 240:1-6.
96. Fansa H, Keilhoff G, Plogmeier K et al. Successful implantation of Schwann cells in acellular muscles. J Reconstr Microsurg 1999; 15:61-65.
97. Farrokh-Siar LA, Rezai KA, Patel SC et al. Cryoprecipitate: An autologous substrate for human fetal retinal pigment epithelium. Curr Eye Res 1999; 19:89-94.
98. Lacy P. Status of islet cell transplantation. Diabetes Rev 1993; 1:76-92.

99. Jos V, Connolly J, Deardon D et al. A simple method for islet isolation from the rabbit pancreas. Transplantation. 1994; 58:390-392.
100. Akita L, Ogawa M, Mandel T. Effect of FK506 and anti-CD4 therapy on fetal pig pancreas xenografts and host lymphoid cells in NOD/Lt, CBA and BALB/c mice. Cell transplant 1994; 3:61-73.
101. Giannarelli R, Marchetti P, Villani G, Coppelli A, Navalesi P et al. Preparation of pure, viable porcine and bovine islets by a simple method. Transplant Proc 1994; 26:630-631.
102. Korbutt G, Elliott J, Ao Z et al. Large scale isolation, growth and function of porcine neonatal islet cells. J Clin Invest 1996; 97:2119-2129.
103. Sun Y, Ma X, Zhou D, Vacek, Sun A. Normalization of diabetes in spontaneously diabetic cynomologus monkeys by xenografts of microencapsulated porcine islets without immunosuppression. J Clin Invest 1996; 98:1417-1422.
104. Auchincloss H Jr, Sachs D. Xenogeneic transplantation. Annu Rev Immunol 1998; 16:433-470.
105. Bach F, Fineberg HV. Call for moratorium on xenotransplants. Nature 1998; 391:326-326.
106. Vogel G. No moratorium on clinical trials. Science 1998; 279:648-648.
107. Ronningen KR, Atrazhev A, Luo L et al. Anti-BSA antibodies do not cross-react with the 69-kDa islet cell autoantigen ICA69. J Autoimmunity 1998; 11:223-231.
108. Dionne KE, Colton CK, Yarmush ML. Effect of hypoxia on insulin secretion by isolated rat and canine islets of Langerhans. Diabetes 1993; 42:12-21.
109. Colas G, Hollands P, Locatelli A et al. The xenotransplantation of goat and human hematopoietic cells to sheep fetuses. Transplantation 1999; 67:984-990.
110. Lohmann T. Limitations of animal models for studying type 1 diabetes. Diabetes Metab Rev 1998; 14(2):192-192.
111. Ao Z, Matayoshi K, Lakey JRT et al. Survival and function of purified islets in the omental pouch site of outbred dogs. Transplantation 1993; 56:524-529.
112. De Vos P, Hillebrands JL, De Haan BJ et al. Efficacy of a prevascularized expanded polytetrafluoroethylene solid support system as a transplantation site for pancreatic islets. Transplantation 1997; 63(6):824-830.
113. Koebe HG, Muhling B, Deglmann CJ et al. Cryopreserved porcine hepatocyte cultures. Chem Biol Interact 1999; 121:99-115.
114. Mullon C. Bioartificial organs may help reduce risk of zoonosis in xenotransplantation. Artif Organs 1999; 23:366-367.

Chapter 2

Synthetic and Semisynthetic Polymers as Vehicles for In Vitro Gene Delivery into Cultured Mammalian Cells

Martin Jordan

Introduction: Impact of Molecular Biology

The impact of molecular biology on everyday life has increased enormously over the last two decades. Medical, pharmaceutical and lately even agricultural applications of "gene technology" have become standard, if sometimes controversially viewed procedures. The feasibility of this "revolution" is based on a few biological facts; most importantly the relationship between DNA, RNA and proteins. DNA carries the information for protein production. Basic units of information are called genes, which typically are DNA sequences of about 1500 base pairs (bp). Usually, one gene carries the information for one protein. While the proteins are highly specific to a species, the genetic code is universal and shared among all living organisms. Therefore, if a human gene is transferred into a bacterium, the bacterium will be able to translate this DNA sequence into the "correct", i.e., human, amino acid sequence (protein). The insertion of foreign genes into bacteria has become a routine laboratory procedure[1] and genetically modified bacteria have been widely used to produce so-called "recombinant" proteins for the pharmaceutical industry. A well-known example is the production of human insulin in *E. coli*.[2]

However, there are limitations to the use of bacteria for the production of proteins, especially of complex proteins from higher organisms. While the genetic code is universal, the machinery for protein processing is not and bacteria lack the enzymes and organelles, which, for example, in mammalian cells are responsible for further processing and modification of the proteins (e.g., glycosylation, disulfide bridge formation, cleavage). Especially in the case of larger proteins, bacteria are often not able to fold the amino acid chain into the correct three-dimensional structure required for "biological activity". Last but not least, the tendency of bacteria to store produced proteins inside the cell in the form of denatured precipitates, so-called inclusion bodies, has been known to considerably reduce the yield of active protein. For this reason, mammalian cells, which have been adapted to propagation in single cell culture, are nowadays used to produce the more complex but also more valuable products of modern biotechnology. Well-known examples are the various CHO cell lines derived from Chinese hamster ovary cells.[3] In order to enable such mammalian cells to produce a desired—human—protein, they too need to be genetically modified. The genetic manipulation of mammalian cells ("transfection") is much more difficult than that of bacteria. Over the last years a

Synthetic Polymers for Biotechnology and Medicine, edited by Ruth Freitag.

number of transfection strategies have been developed, amongst the methods that utilize (semi-)synthetic polymers. A controllable and successful transfection strategy is not only the basis for the production of recombinant proteins, but even more so for gene therapy. Considerable attention has therefore been paid to the development of synthetic polymers as vehicles for gene delivery. This chapter will focus on the current state of knowledge in regard to the requirements for putative transfection vehicles, but also will summarize and compare the various applications of such systems.

Areas in Need of Efficient Gene Delivery

Today an enormous amount of genetic information is available from databases, which are continuously fed by worldwide genome sequencing programs (e.g., www.sanger.ac.uk, www.ncbi.nlm.nih.gov). Every day, the human genome-sequencing program alone provides new information about human genes with potential therapeutic value. Conservative estimates are that, by the year 2001, all of the approximately 100,000 human genes will have been sequenced. On a diagnostic level, this will allow detecting "genetic defects" and also a prediction as to which amino acid in a given protein is concerned. However, unless the change in amino acids is meaningless or the malformed protein can be replaced, this information has limited therapeutic impact since curing the DNA defect is at present not possible. Another aspect concerns the large number of genes with unknown function. Since it is not possible to predict the three-dimensional structure of a protein, let alone its biological function (interactions with other biological substances), from its amino acid sequence, the only way to "mine" the genetic information consists in a laborious transfection of a mammalian cell with the gene in question to enable said cell to produce the protein. Subsequently this allows studying the activity of the protein either directly within the cell or in vitro once enough of the material has been produced for further characterization.

The problem of quickly producing a certain amount of protein for further characterization and study is a major bottleneck in several areas of the life sciences and the related bioindustry. The list of sequenced genes for which the function of the corresponding protein is poorly understood is long. In addition, it is fairly easy to mutate genes in vitro, so a variety of new proteins can be encoded, some of which might have considerable therapeutic value. In contrast to the quick generation of new genes, the establishing of a stable recombinant production cell line requires at least a year for transfection, screening/amplification and scale up due to the difficulties of inserting the gene stably into a transcriptionally active region of the cell's chromosomal DNA. Recently, a much faster method—transient transfection—has been discussed as a means to produce quickly (within days) milligrams of a given protein. In this case, the foreign DNA is not inserted into the genome of the cell (see below). The method, which until recently was only used for the production of smaller amounts of proteins through-out, had been shown to be compatible with at least the 1 L scale.[4] If transfections could be established at the 100 liter scale or more, gram amounts of any protein could be produced within days.[5] Screening of putative biopharmaceuticals but also basic research would profit enormously. Such large-scale transfections have not yet been achieved.

Gene therapy is another domain where efficient transfer of genes is essential. Many severe human diseases are caused by a genetic defect leading to the mal- or over-/ under-expression of the corresponding protein. Patients could be permanently cured if the missing genes could be transferred in a functional form into the concerned organs. Delivery of genes to specific tissues could become the most efficient medical treatment in the future, but for obvious reasons, the establishment of a very safe and well-controlled method for gene delivery is imperative.

The DNA Molecule

The structure of the large DNA molecule, which was known to be the main material of the chromosomes, remained a mystery until Watson and Crick proposed the double helix structure in 1953.[6] Chromosomal (genomic) DNA consists of two complementary polyanionic chains made up of long sequences of four different nucleic bases. Since the four bases are complementary, the double stranded DNA molecule is capable of exact self-replication from either strand. The chemical structure of DNA is shown in Fig. 2.1. The diameter of the double helix is about 2 nm, while the length of the DNA polymer can be enormous, i.e., several centimeters in a putatively "stretched out"-state. Table 2.1 gives some comparisons for the size of DNA from different species.

In a typical human cell, DNA molecules with a total length of 1 meter have to be packed into a nucleus of about 5 µm in diameter. The compaction is mediated by the so-called nucleosomes, which contribute about 50% of the total mass of the chromosomes. Nucleosomes are formed by 4 to 5 different types of histones; small, basic proteins with a high proportion of positively charged amino acids (25% lysine or arginine). Histones, which bind tightly to the sugar-phosphate backbone of DNA, also have important regulatory functions. Compacted DNA is not active, meaning it can be neither replicated nor transcribed into RNA and finally into proteins. The histones control the compaction and the compaction-reversal through a regulated process that is gene or sequence specific. The exact biochemical basis of this regulation strategy is still unknown, but the essential role of histones in life is supported by the fact that their amino acid sequence is among the best conserved throughout evolution. Apparently, even minor changes in the histone structure have dire consequences for the organism in question.

Plasmid DNA is an independent type of DNA, which occurs naturally in many microorganisms in addition to the genomic DNA of the respective organism. Plasmids are comparatively small (typically 5–10 kb[a]), circular DNA molecules that can multiply independently from the genomic (chromosomal) DNA. They occur naturally in the supercoiled (major percentage) and the open circular form (see Fig. 2.2). Linearized fragments of plasmid DNA can be obtained by "digestion" of the plasmid with restriction endonucleases, i.e., enzymes that cut the DNA at specific base pair sequences.[b]

For various reasons, plasmid molecules are the preferred tools for genetic engineering. Plasmids can easily be amplified in bacteria. They are separated from the larger chromosomal bacterial DNA by a denaturation/renaturation process, where the chromosomal DNA forms an insoluble precipitate, because it renatures more slowly. Purified plasmids can be transferred into eukaryotic cells either in their natural, supercoiled form or as linearized molecules.

Barriers to Efficient Gene Transfer

DNA, the common carrier of the genetic information for all living entities on this planet, is omnipresent and we are daily exposed to large quantities of foreign DNA (e.g., by food or bacterial infections). Under these circumstances, nature had to provide powerful barriers against the spontaneous insertion of foreign DNA sequences into the genomic DNA of cells. Barriers are the plasma membrane of the cell, the envelope of the cell's nucleus, but also the possibility for DNA degradation in lysosomes and the cytoplasm (see Fig. 2.3). These protective mechanisms work rather well and even under optimized conditions it is by no means easy to genetically modify an eukaryotic cell (the terminus usually employed for this modification is to "transfect" the cell). However, the necessity to transfect cells for research purposes, the discovery of new and efficient reporter systems to verify the success of a transfection experiment (luciferase, green fluorescent protein) as well as the availability of powerful transfection

---

[a]Kb: kilo base pair
[b]Most restriction nucleases recognize a sequence of six or eight base pairs

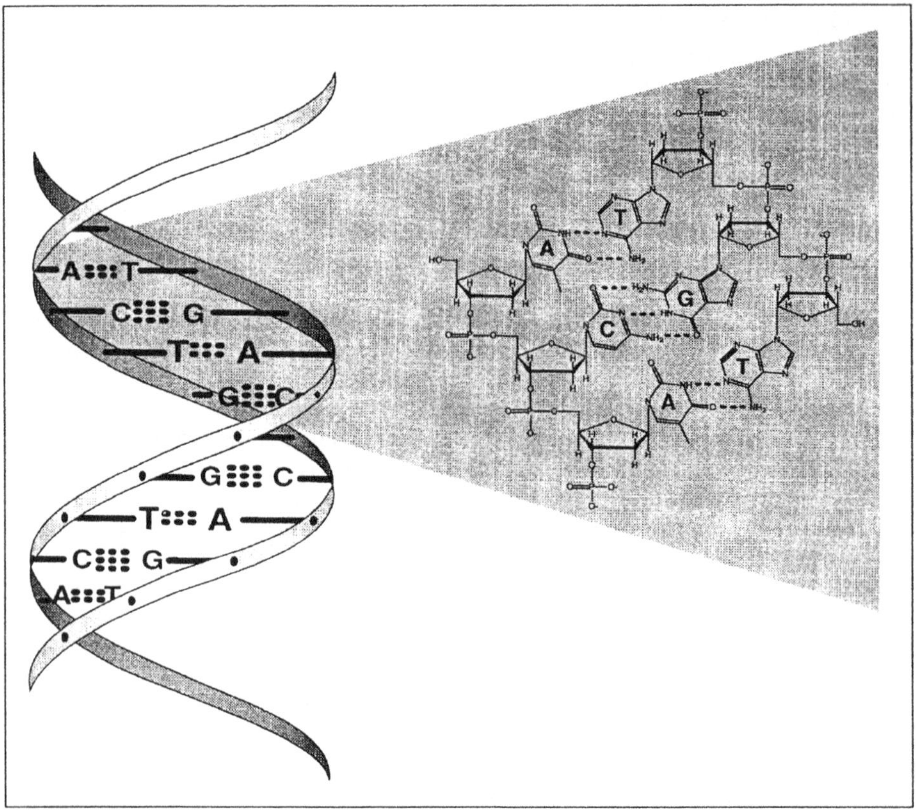

Fig. 2.1. The DNA double helix: The Watson-Crick structure and the chemical composition.

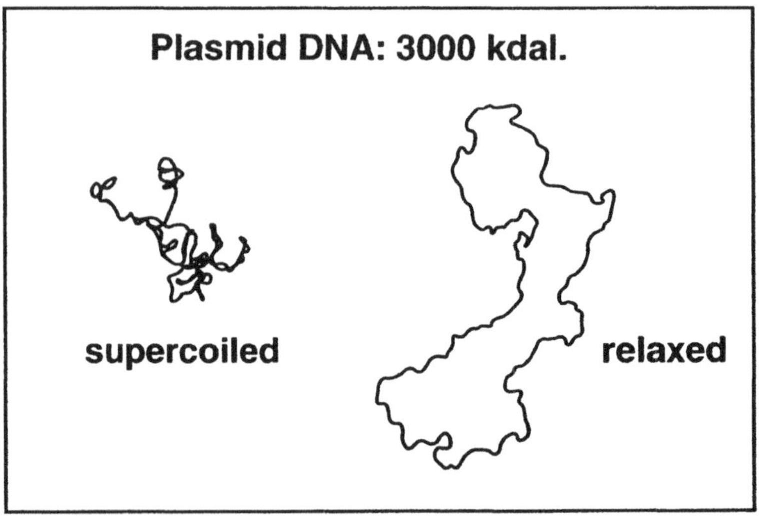

Fig. 2.2. Comparison of size and structure of supercoiled versus relaxed circular plasmid DNA.

Table 2.1. Size of genomic DNA

Organism	Number of basepairs (kb)	Length (μm)	Mass (kg/mol)
Virus (SV40)	5	1.7	3,300
Bacterium (*E. coli*)	4,000	1,360	2,640,000
Yeast	13,500	4,600	8,910,000
Drosophila	165,000	56,000	108,900,000
Human	2,900,000	990,000	1,910,000,000

reagents have spurred research in the area for many years. Several methods to transfer genes into cells have been developed during the last 30 years. However, considerable efforts to develop new techniques or to improve the efficiency of old ones are still being made.

Transfection reagents help to overcome the natural barriers to gene transfer by various strategies. The steps involved in the transfer of a "gene" from the outside into the genome of the cell comprise (Fig. 2.3):

1. compaction of the DNA,
2. attachment to the cell surface,
3. transport into the cytoplasm,
4. import into the nucleus and
5. insertion into the chromosomal DNA.

The mechanism by which a certain barrier is overcome is an important feature of the respective transfection reagent. In order to elucidate the difficulties in optimizing the genetic engineering of mammalian cells, the major steps of transfection as well as putative agents for reaching this goal will be discussed in detail in the following sections. The mechanisms for many of the above-mentioned five steps of transfection are still under discussion. This is especially the case for the later steps taking place inside the cell, i.e., transport into the cell and most importantly into the nucleus. The earlier stages of compaction and interaction with the cell surface are better understood. This has important consequences for our current ability to engineer transfection agents and procedures. It should be noted that man-made transfection procedures are still orders of magnitude less efficient than nature's transfection agents, the viruses are. One to five infectious particles, i.e., viruses, per cell are sufficient in that case, compared to the 10^5–10^6 plasmid molecules needed in most nonviral transfection methods.

Compaction of DNA

Pure ("naked") DNA has little chance to enter a cell. DNA is a huge, negatively charged and hence highly hydrophilic molecule. Cells are surrounded by a hydrophobic plasma membrane and, in addition, bear a negative surface charge. The plasma membrane contains several highly selective transporter units, which allow for the well-controlled introduction and excretion of certain molecules. Foreign DNA is normally not amongst the molecules allowed to enter the cell.

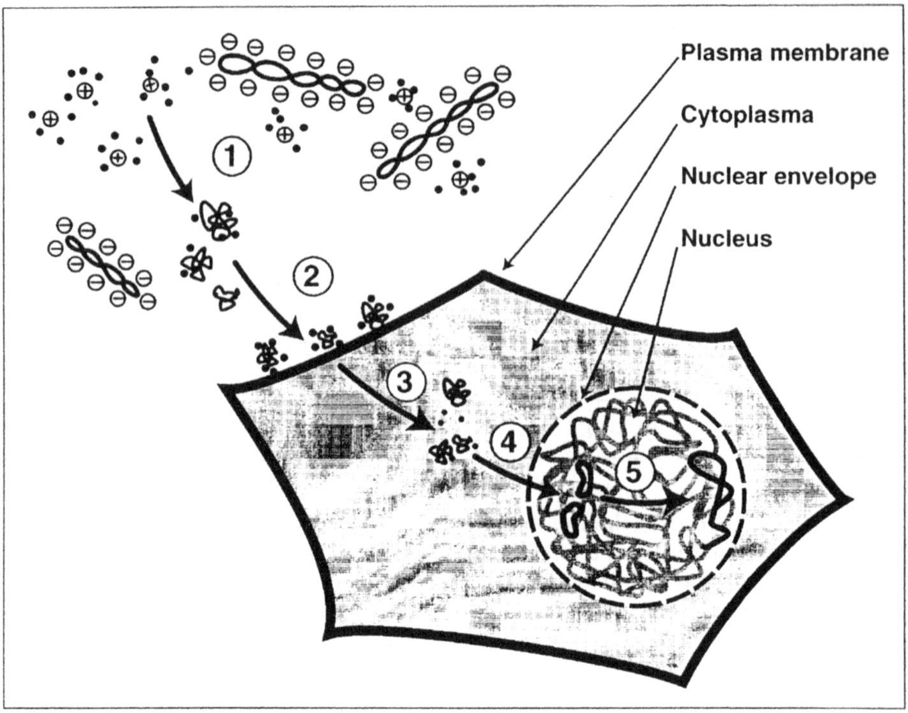

Fig. 2.3. Essential steps for transfection: 1) compaction of DNA, 2) surface attachment, 3) transport of DNA into the cytoplasm, 4) import of DNA into the nucleus, and 5) insertion of the DNA into the chromosome.

The first and best-understood step of transfection is therefore the necessity for "compaction" of the large, negatively charged DNA molecule. A suitable compacting agent is a positively charged molecule able to interact with the DNA and to neutralize or even overcompensate the negative charges. During compaction, the DNA forms stable complexes with the compaction agent, which either stay in solution or form a precipitate. In a typical transfection experiment, the complexes are formed in a reaction mixture containing the given amounts of purified DNA as well as the compaction agent under defined pH and salt conditions. The complex formation occurs spontaneously upon mixing. Within the next 30 minutes the complexes are added to the target cells. Usually, cells are exposed for several hours to the complexed DNA. Subsequently, the medium is exchanged in order to minimize possible toxic effects.

Two groups of molecules are currently investigated as compaction agents: cationic lipids and cationic polymers. Protonated amino groups provide the required positive charges in both cases. Amino groups are also found in some of the naturally occurring compaction agents such as spermine and spermidine. They are clearly the group of choice, since they allow the generation of a positive charge at physiological (neutral) pH. In addition, eukaryotic cells developed over eons of evolution special proteins (nucleosomes) with a high affinity to DNA, which also can complex DNA. The structure of these nucleosomes may in the future inspire the design of novel compaction agents. Prominent representatives are histones or protamines, naturally occurring ubiquitous DNA binding (compacting) proteins.

Cationic lipids (Fig. 2.4a) are usually fairly small molecules, which mimic the structure of the cell's plasma membrane and hence facilitate the passage of DNA into the cell by increasing the solubility of the DNA in the plasma membrane. These molecules consist of a hydrophobic

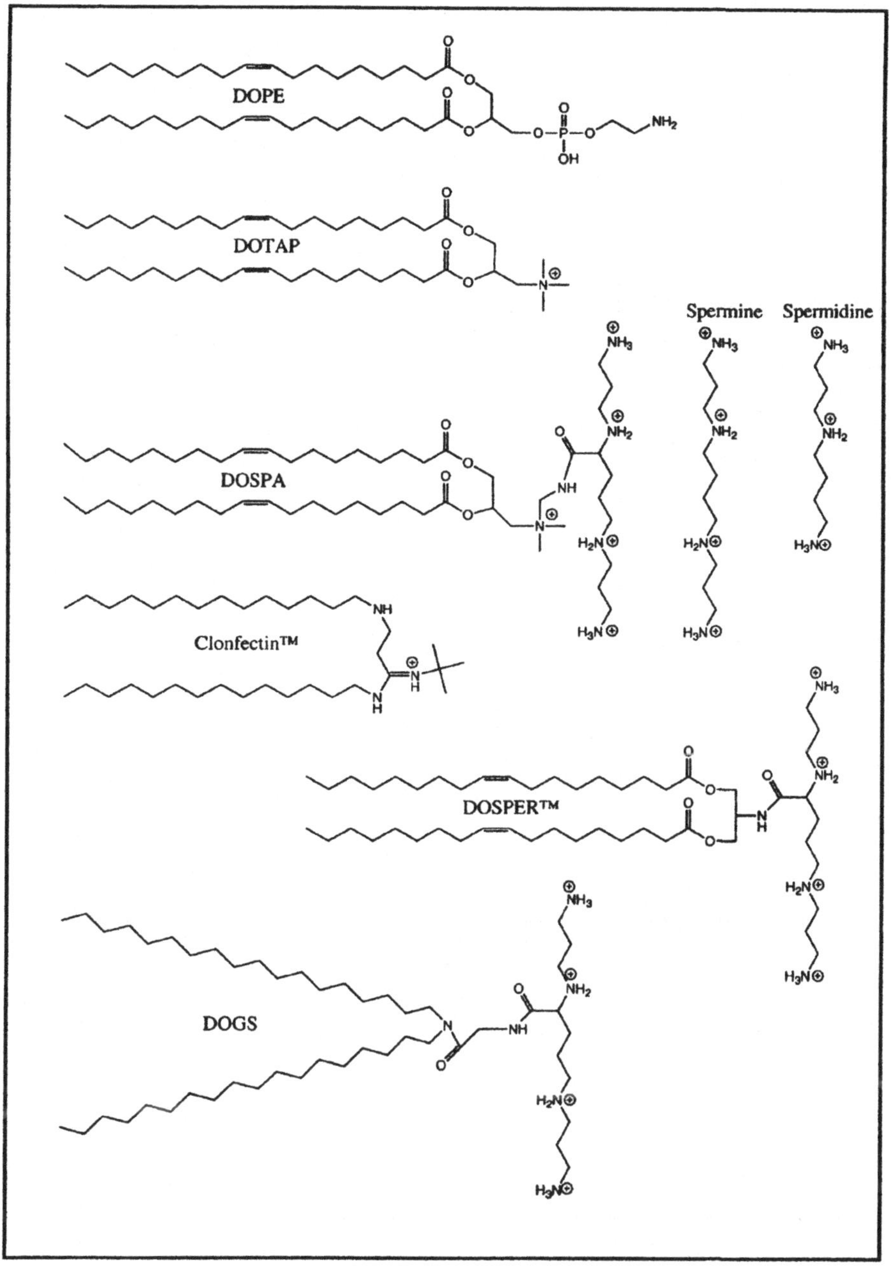

Fig. 2.4a. Examples of cationic lipids.

(hydrocarbon) tail and a positively charged head-group. The hydrophobic tail promotes in aqueous solutions self-aggregation into larger structures[c] (micelles, double layers) capable of interaction or even fusion with the cellular membrane.

[c]Most cationic lipids are commercially available as liposomes, small spherical vesicles with a lipid bilayer

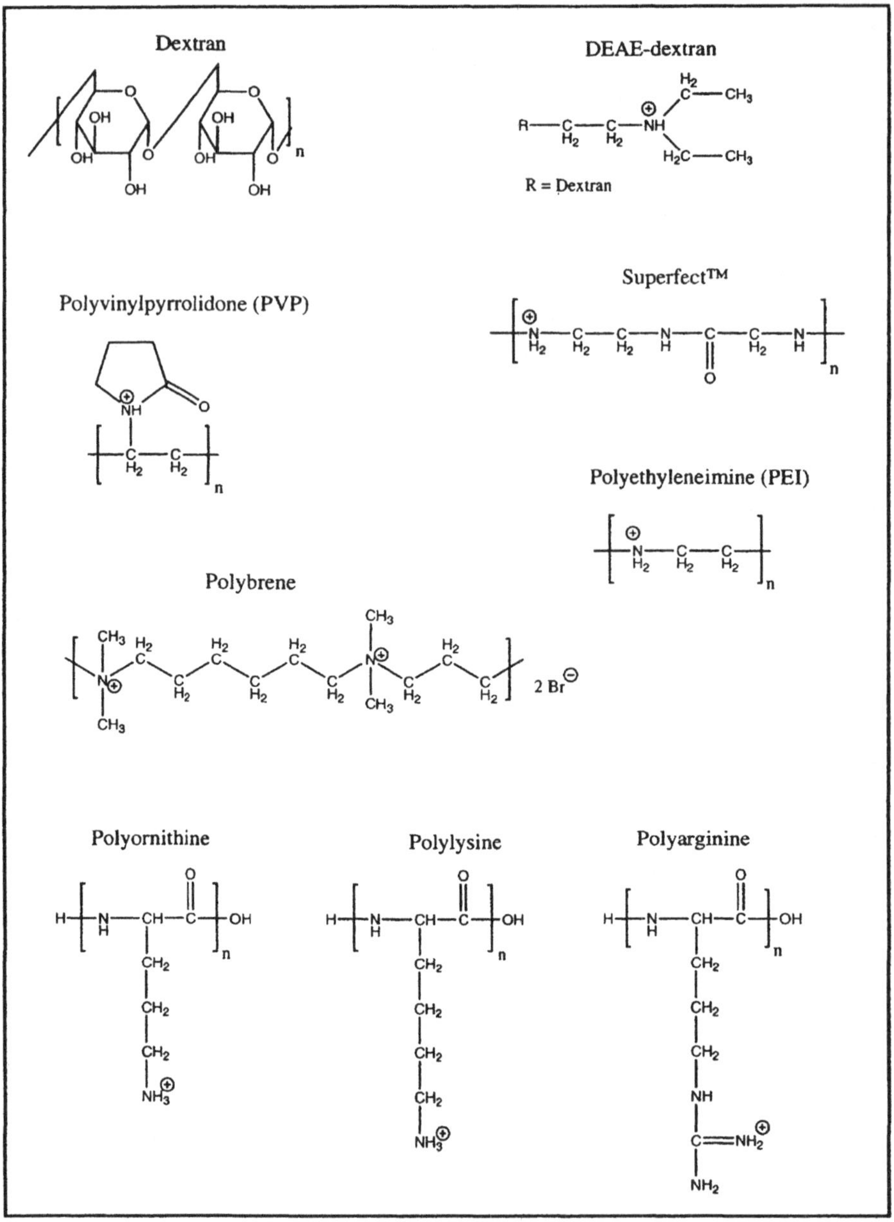

Fig. 2.4b. Examples of cationic polymers.

The cationic polymers (such as polyethyleneimine, polyvinyl pyrrolidone) commonly used for transfection (Fig. 2.4b) are fairly large molecules (up to 1,000,000 g/mol). They are soluble in water at neutral pH due to their positive charges. Linear as well as branched molecules are employed for transfection. In contrast to the cationic lipids, which usually were developed as dedicated transfection reagents, most cationic polymers have been developed for other

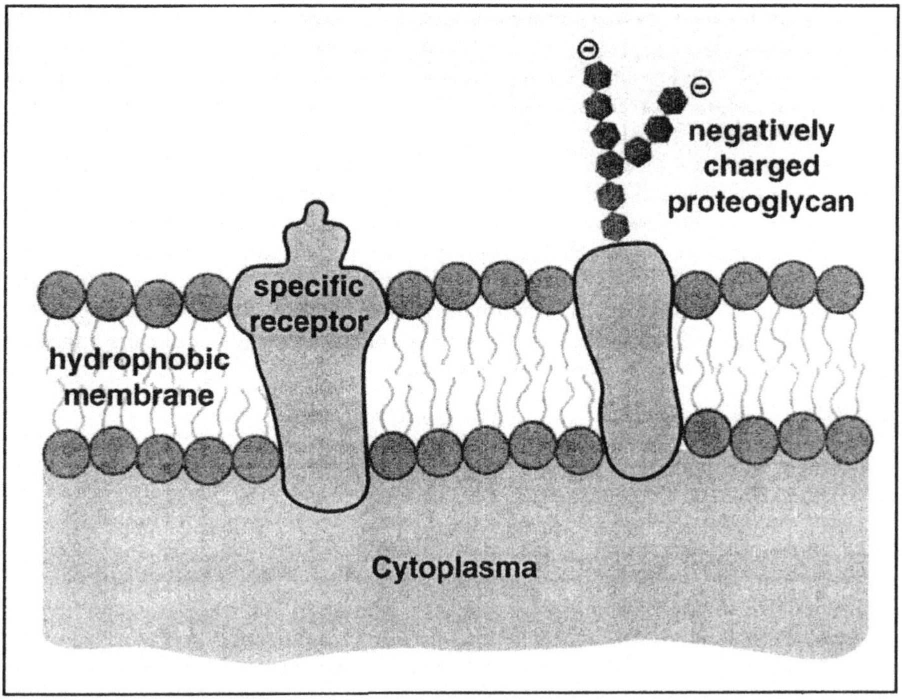

Fig. 2.5. Cell surface with possible targets for interaction: the cell membrane includes a large number of proteins having various functions such as structure, transport of molecules, and signaling; cell types differ by quantitative and qualitative content of membrane proteins.

applications and purposes. They are therefore available from several suppliers in a wide variety of purity and chemical homogeneity.

Attachment to the Cell Surface

If a foreign DNA sequence is to be introduced into a cell, it is obviously necessary that the two meet, i.e., that the compacted DNA somehow attaches to the cell surface within and for a reasonable amount of time. The basic structure of the cell membrane is given in Figure 2.5. Cell membranes consist of a lipid bilayer into which a number of complex (glyco)protein molecules are inserted or anchored. The dominant mechanism for interaction between the DNA complex and the negatively charged cell surface are electrostatic forces. The negative surface charge is in many cases provided by proteoglycan molecules carrying anionic sulfate groups, which are present on the surface of many cell types. Positively charged complexes may attach themselves to the cell surface via these molecules.[7] The importance of this type of interaction to the success of a transfection has been demonstrated by the following experiment. It has been shown, that DNA charged cationic liposomes fail to transfect so-called Raji cells, which are proteoglycan negative, but transfect genetically modified, proteoglycan positive (syndecan-1), cells of the same cell line with good efficiency.[8]

Electrostatic interaction with the proteoglycans, however, is not the only possibility for interaction between a DNA-carrying transfection agent and a cell surface. Many membrane proteins expose binding sites (receptors) for certain biochemical messenger molecules (ligands). In general, such receptor proteins control the specific uptake of molecules and make the cell sensitive to hormones and other signal molecules. This natural mechanism can be subverted for DNA transfer. The receptor ligands can be used to increase transfection efficiency in general or they can be used to target the transfection complex to a specific cell or tissue type by evoking an interaction between the transfection complex and a cell-specific receptor. Targeting can, for example, be achieved by introducing ligands such as, insulin,[9] transferrin,[10] lactose,[11] galactose,[12] mannose,[13] folate,[14] poly(acrylic acid)[15] or specific monoclonal antibodies[10] or antibody fragments[16] into the transfection complex. This addition has been shown to dramatically increase the efficiency of transfections with agents such as poly(lysine) or poly(ethyleneimine) for certain cell lines, which otherwise were difficult to transfect. It seems that the improvement is due to the ligand's ability to subsequently induce receptor-mediated uptake of the DNA into the cell (endocytosis, see below). In addition, receptor mediated transfection can be blocked (controlled) if necessary by complementing the cell culture medium during the transfection with an excess of the free ligand.[14]

Transport into the Cell

Two basic mechanisms are assumed to contribute to the transport of the DNA into the cell. These are: a) an active, energy dependent uptake of the transfection complexes by a process called endocytosis[d] (see below) or b) "passive" membrane fusion and release of DNA into the cytoplasm. The compaction agent used in the first step largely determines which mechanism is more important in a given case. For polycationic molecules, a direct interaction (fusion) with the hydrophobic membranes is not likely. The most likely way for them to enter the cell would be by endocytosis. Cationic lipids, on the other hand, can potentially interact and fuse with the membrane. Experiments with synthetic membranes have demonstrated the fusogenic ability of liposomes formed by cationic lipids,[17] but convincing data that this mechanism is also operative during transfection of living cells are still lacking. Other reports seem to indicate that liposomes also preferably enter the cell by endocytosis.[18-20]

Endocytosis is a process by which cells take up extracellular molecules such as cholesterol via a receptor-mediated mechanism (Fig. 2.6). Cholesterol, insoluble in aqueous solutions, naturally occurs in association with the so-called low-density lipoproteins (LDL). The uptake of cholesterol by the cells depends on receptors specific for LDL. In a first step the ligands bind to the receptor.[e] Receptors occupied with ligands form clusters and induce the formation of a clathrin-coated pit. Clathrin induces the expansion of the pit. Such pits can subsequently enter the cell as a membrane-bound vesicle containing the ligand/cholesterol-complex. Inside the cell, the vesicle rapidly loses its clathrin coat. Vesicles containing receptor bound ligands undergo further changes. Protons are actively imported into the vesicle leading to a drop in pH from the physiological values of 7 to about 5. Under these mildly acidic pH conditions, receptor and ligand dissociate. Receptors are then recycled back to the membrane with the aid of a sorting vesicle.[f] The ligand/cholesterol-complexes stay within the vesicle and are transported towards the so-called lysosome, an even more acidic vesicle containing digestive enzymes. In the case of the ligand LDL, the ligand/cholesterol complex is digested inside the lysosome into amino acids, cholesterol and fatty acids.

Receptor mediated endocytosis may easily be exploited for DNA transfer into a cell, but, if DNA ends up in a lysozyme, it will be degraded. In order to succeed with gene transfer, the DNA needs to escape the endosome before it is digested by lysosomal nucleases. This is

---

[d]Phagocytosis and pinocytosis, which are not very different from endocytosis, will not be described in this article.

[e]Each cell line has a characteristic number of receptors for LDL, insulin, transferrin etc.

[f]Recycling of receptors is essential for the cells: e.g., cultured fibroblasts can internalize regularly 50% of their cell surface proteins and phospholipids per hour.

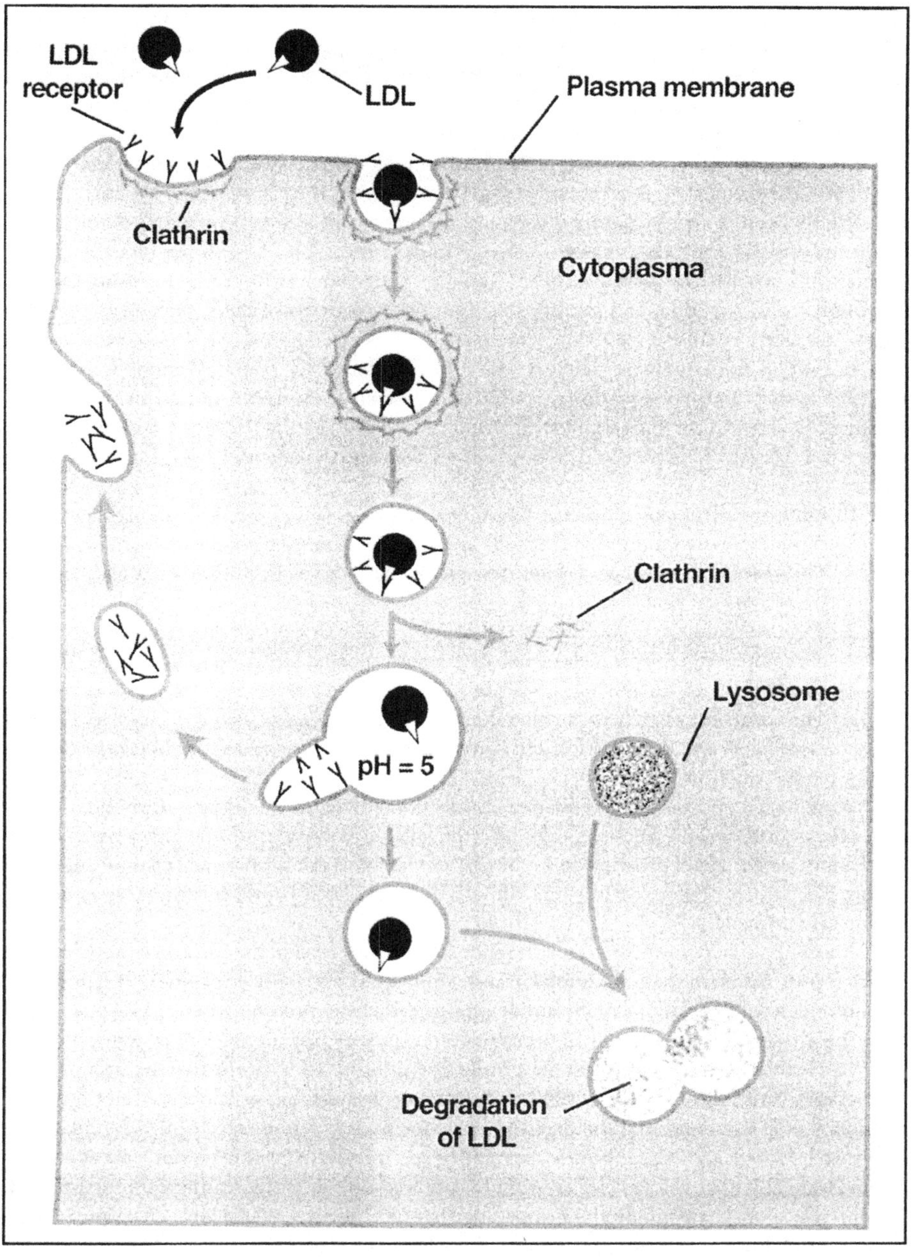

Fig. 2.6. Receptor-mediated endocytosis: example of low density lipoprotein receptor with the following steps: receptor-mediated pit formation, formation of vesicles, pH drop inside the vesicle, fusion with a sorting vesicle, recycling of a receptor, fusion with lysosome and digestion.

possible, as demonstrated by a number of infectious viruses, which use endocytosis for the efficient transfer of their genetic material into certain target cells. Such viruses have special capsid proteins that allow them to escape the early endosome. The signal for their escape is triggered by the drop in pH.[21,22] As soon as the pH in the endosome starts to decrease, the capsid proteins undergo a conformational change that enables them to fuse with the membrane of the early endosome. The result is a disruption of the vesicle and the release of the virion into the cytoplasm. A synthetic peptide derived from the capsid of the hepatitis A virus has recently been shown to mimic this endosome escape induced by low pH.[23] Another, less efficient, way to escape the lysosome consists in the utilization of lysosome blocking agents such as chloroquine[24] or—even simpler—in an osmotic shock enforced by exposing the cells to nontoxic and nonionic compounds but osmotically active molecules such as glycerol[25,26] and DMSO.[27]

In spite of this convincing picture, the role of endocytosis for transfections is at present not fully understood. For one thing, much less is known about DNA uptake in the artificial situation of transfection than about the natural uptake of clinically relevant substances such as cholesterol. However, many observations support the theory that endocytosis is a key factor in transfection.

1. Electron microscopic studies of DNA uptake during transfection showed that transfection complexes were found inside the cells in vesicles surrounded by biological membranes.[18,28] The pH of such vehicles was found to be acidic, a characteristic feature of late endosomes or lysosomes.
2. The incorporation of inactivated adenovirus particles or parts of viral capsid proteins into the compacted DNA complexes enhances transfection efficiency up to three orders of magnitude.[29]
3. The treatment of cells with chloroquine, DMSO or glycerol strongly improves levels of expression and transfection efficiency, presumably due to partial blockage of lysosome function.

More recent approaches to enhance transfection efficiency use fusogenic peptides to improve the performance.[30] Such peptides help the DNA to escape lysosomic digestion in a similar manner as the capsid proteins do in the case of viruses. A conformational change is induced at low pH, which triggers a fusogenic activity. Typically such peptides contain virus derived amino acid sequences.

Another putative route to escape the endosome is proposed in the "proton sponge" theory, which is postulated for the polycationic transfection agent poly(ethyleneimine), PEI, and similar molecules.[31] PEI contains many amide groups, which are protonated at a pH below 7. PEI thus constitutes a gigantic buffer molecule (proton sponge), the presence of which would prevent a pH drop in the early endosome simply by capturing the protons that are pumped into the vesicles. Since the decrease of pH is prevented or delayed, the vesicle would not fuse with lysosomal vesicles containing the digestive enzymes, and therefore rapid DNA degradation is prevented. How exactly the DNA escapes subsequently from the vesicles is not clear at present, but it is possible that the vesicles are simply ruptured due to an increasing osmotic pressure. The proton sponge effect may also be operative in one of the oldest and most efficient transfection methods known to molecular biology, the calcium phosphate technique.[32] In this case, the DNA is coprecipitated with calcium phosphate, presumably due to an involvement of the phosphate groups of the DNA in the crystals. The precipitate particles are taken up by the cells via endocytosis. Calcium phosphate is not stable at acidic pH and as the pH inside the endosome starts to drop below 7, the complexes dissolve and release the buffering ion PO_4^{-3}, which stabilizes the pH at a more physiological level for some time.

Entering the Nucleus

The DNA is now inside the cell and more precisely in the cytoplasm. It still has to reach its final destination, the cell nucleus, i.e., pass the nuclear membrane. Microinjection experiments demonstrate that this membrane indeed constitutes a barrier.[33] By microinjection, DNA may be injected into a cell either into the cytoplasm or directly into the nucleus. In order to achieve the same result, the amount of DNA injected into the cytoplasm has to be at least ten times as high as the amount injected directly into the nucleus.

Two mechanisms to overcome the nuclear membrane are postulated. DNA may be actively transported into the nucleus through pores in the membrane, which normally enable the import of histones or other nuclear proteins into the nucleus. This active import into the nucleus can presumably be activated either by the DNA sequence itself[34] or by coupling the DNA/compaction agent-complex with proteins or peptides bearing the nuclear targeting sequence. Targeting peptides can, for example, be derived from histones,[35] viruses or transcription factors.[36-38] Alternatively, DNA may enter passively during mitosis (cell division), when the nuclear membrane is known to disappear completely. However, mitosis lasts only a few minutes; a small time window when one considers that the entire cell cycle takes between 12 and 30 hours. This would mean that DNA has to "wait" for several hours in the cytoplasm for the next mitosis to occur. With a DNA half-life of 1-2 hours the chances for successful transfection would decrease exponentially over time.[39] That mitosis does have some influence on transfection efficiency was demonstrated by experiments where the cell cycles of all cells in a given culture were synchronized. When such synchronized cells were transfected, best results were found when mitosis occurred a few hours after the addition of the transfection complexes,[40,41] i.e., when the DNA had had enough time to leave the endosome but had not been degraded to a significant degree. However, were mitosis the only possibility for DNA to enter the cell, then nondividing cells would not be transfectable at all, which clearly is not the case.[42]

Transient Versus Stable Expression

Once the DNA has found its way into the nucleus, it serves as a template for the corresponding protein, since all the enzymes relevant for transcription are present in the nucleus. However, foreign DNA, which is not integrated into the chromosomes will not be replicated during the cell cycle and therefore will be lost after a few cell divisions. The gene expression of nonintegrated foreign DNA is therefore "transient." In contrast to transient expression, stable expression requires the integration of the foreign DNA into a chromosome. Then it will be passed on to the daughter cells during the cell cycle together with the rest of the chromosome. Only a few of the DNA molecules that arrive in the nucleus will eventually be integrated into the chromosomes. With the exception of specific systems that promote sequence specific integration with the help of targeting and recombination promoting proteins, little is known about agents or procedures capable of improving the efficiency of random integration of foreign DNA into the genomic DNA.

Nonviral Transfection Methods

The simplest way to expose cells to foreign DNA is to mix the "naked" plasmid DNA with the cells.[43] As was already discussed, however, the success rate to be expected in terms of stable or even transient transfection is extremely low in this case. The DNA uptake is inefficient and the expression of the respective proteins can only be detected with the most sensitive assays (luciferase, immune response). The method is more promising if used in combination with electroporation. In this case an electrical pulse[g] is used to create temporary pores in the cell membrane through which the DNA can enter. While electroporation is fast, it needs special

[g] Typical pulse time $t = 10 - 40$ msec.

equipment and cells have to be placed into the electroporation cuvettes. In addition under optimized transfection conditions, up to 50% of the cells are killed by the electrical pulse.[44] Electroporation is obviously not suited to application for in vivo gene therapy or at large scale.

Chemical methods are currently the most promising alternative to virus-mediated transfection of mammalian cells. Their efficiency is still below that of the virus-based methods, but questions of toxicity and the remaining risk factors (viral infections) are more favorable in their case. It remains to be seen if a further improvement of the transfection agent is possible, e.g., by taking some of the above-mentioned biological and biochemical barriers into account. The penultimate goal would be a completely nonviral transfection agent, which is nevertheless equal in efficiency to today's virus-based ones.

The ideal transfection reagent should be a charged molecule that can form complexes with DNA in aqueous solutions and thereby compact the DNA. The agent needs to be sufficiently soluble and stable in aqueous solution at physiological pH. However, the complexes should not be so stable as to prevent the release of the DNA within the nucleus of the cell, since this would interfere with transcription of the gene. Minimal toxicity towards cells is another critical feature. The strong affinity to DNA should result in spontaneous and reproducible formation of complexes upon mixing at room temperature. They should have a maximal size of several hundred nanometers. Otherwise the complex will not be taken up by endocytosis. In addition, complexes should be insensitive to isotonic salt concentrations and slightly positively charged, the latter to actively attract them to the negatively charged cell surface. Typical components of cell culture media such as amino acids, vitamins, salts and trace elements should not influence the complex formation or interact with the formed complexes. The same is true for the more complex media additives such as proteins (insulin, transferrin amongst others), lipid mixtures or fetal calf serum. Last but not least, the "user friendliness" of the transfection protocol and the price (reflecting the difficulty of manufacturing of the agents) also play an important role, especially if the method is to be used on a large scale. Below some of the more typical chemical transfection agents are discussed in view of these criteria.

Synthetic Polycations

A number of cationic polymers have been shown to be powerful transfection reagents. It is evident that fairly large polycationic molecules are needed to interact sufficiently with the plasmid DNA, typically a molecule with a size of 5000 basepairs or more. The first polymer ever described to enable the transfer of viral DNA into mammalian cells was diethyl aminoethyl-dextran (DEAE-dextran, molecular mass 500,000 g/mol) in 1968.[45] DEAE-dextran is still used today for that purpose in many laboratories worldwide. Partially responsible for this success may be the fact that DEAE-dextran is known to bind both cells and DNA and thus actively brings them together. This carries the plasmid DNA through the first two steps of transfection, i.e., compaction and attachment to the cell surface. For the actual transfection experiments, diluted DEAE-dextran is mixed with diluted DNA, since without dilution, an almost insoluble precipitate forms. Once the complexes have formed, they are added to the cells in transfection buffer. After an incubation of 1 hour a suitable growth medium is added. Alternatively, the cells are first treated with DEAE-dextran. Unbound DEAE-Dextran is subsequently removed and the DNA is added to the dextran-covered cells. The mixture is then incubated for up to 1 hour before cell culture medium is added. A major drawback of the use of DEAE-dextran for transfection is the fact that the complex is quite fragile. Certain components of the growth medium are known to interfere seriously with complex formation. Growth medium must therefore be strictly avoided. Only a few cell lines can be transfected directly using DEAE-dextran. Even then it is often necessary to treat the cells either with chloroquine during transfection or to apply a short DMSO shock afterwards in order to obtain good transfection efficiencies.[46]

Poly-L-lysine and poly-L-ornithine are two molecules (polyamino acids), which also have a long history of being used as successful transfection agents.[47,48] Poly-L-ornithine might be slightly more efficient than poly-L-lysine for transfections, but it also has an higher cell toxicity, especially before it is complexed with DNA. Poly-L-lysine is less toxic and commonly used in cell culture technology to immobilize cells onto surfaces. Lysine is a positively charged, natural amino acid that is found in proteins including the histone proteins, where together with arginine it contributes a major part of the building blocks. The structural similarities between lysine, ornithine and arginine (Fig. 2.4b) have led to an investigation of the latter as transfection agent. However, neither poly-L-arginine nor poly-L-lysine/poly-L-arginine copolymers were efficient transfection reagents.

Poly(ethyleneimine), PEI, was only recently discovered as a very potent transfection agent.[31,49] Since it can be bought in bulk quantities, it is much cheaper than most other reagents discussed so far. It reacts very strongly with DNA; complexes will form even if DNA and PEI are simply added successively to cells in a serum-free cell culture medium. Complexes are also quite stable in the presence of serum and transfection can be done directly in cell culture growth media. Transfection by DNA/PEI-complexes is possible with cells grown on a solid support (adherent cells) as well as with cells grown in suspension. Large scale transient transfections for the rapid production of proteins up to the scale of 20 liter yielded several mg of protein within days using human embryo kidney 293 cells as production organisms.[50]

Currently, a linear ultrapure PEI (molecular mass 22 g/mol) is commercially available as a transfection reagent. However, at least in our hands, such a preparation was not superior to the branched bulk PEI ordered from a standard supplier of chemicals (25 g/mol PEI from Aldrich Chemical Company Inc., Milwaukee, USA).

The advantage of PEI over many other polycationic transfection agents consists in the buffering capacity of the amino groups ($pK \approx 6$) of the polymer's backbone. This feature helps the DNA to escape the lysosome before being degraded since PEI acts as a proton sponge. It is interesting to note that most potent transfection agents bear amino groups with a pK slightly below neutral pH. They are probably also capable of some buffering activity. However, PEI has the highest buffer capacity of all transfection agents used so far, since 33% of its mass consists of protonable nitrogen atoms. In comparison, the corresponding mass fractions are only 11% in the case of poly-L-lysine, 12% in the case of poly-L-ornithine and 3% in the case of DEAE-dextran.

Polybrene, Figure 2.4b, is a synthetic polymer, known to retrovirologists for its ability to augment the infectivity of viral particles.[51,52] Recently, it has been also described as an effective transfection reagent.[53] It is reported that positive, i.e., transfected, cells are already obtained with only nanograms of DNA, in comparison to the micrograms necessary with most other methods.[54] Polybrene-DNA complexes uniformly coat the target cells and are then internalized. However, the method is only effective if it is combined with a DMSO shock and has no particular advantage over the use of PEI.

Up to now, linear and low-branched synthetic polymers have been discussed as transfection agents. In addition, at least one dendritic transfection reagent is commercially available as SuperFect™. Chemically speaking, SuperFect™[55] is a polyamidoamine starburst polymer with a diameter of about 7 nm and a molecular mass of 35,000 g/mol. Heat-activated dendrimers swell to 20-30 nm. Concomitantly, the potential as a transfection reagent is increased by about three orders of magnitude. Activated dendrimers have a strong affinity to DNA and compact DNA molecules into small particles. Although the DNA is adsorbed at the surface of the dendrimer, it becomes completely protected from nucleases. Complexes are formed within 5 minutes and they can be frozen. Once formed, the complexes do not lose activity in the presence of serum. For the actual transfection, the complexes are simply added to the cells and incubated for a few hours. Afterwards remaining complexes should be removed in order to avoid toxic effects.

A number of the above-mentioned reagents have also been used for in vivo gene therapy, i.e., the genetic modification of cells within a living (human) being. DNA/poly-L-lysine complexes were, for example, injected into the bloodstream and subsequently distributed all over the body. Some target specificity could be achieved by coupling the transfection agent to a ligand, which in turn was specific for a target cell receptor, for example the transferrin receptor.[56] This enhanced the uptake of the complexes into cells bearing this particular receptor on the surface. A problem associated with poly-L-lysine as transfection agent is the overall transfection performance. Although many cells take up the complexes very efficiently, the transfection outcome is still relatively poor, mainly because most of the DNA is degraded in the lysosome. Adding psoralen-inactivated adenovirus particles dramatically improved the transfection efficiency in such cases, presumably due to the disruption of the lysosomal membrane by the virus. By using this mechanism, the transfection efficiency of poly-L-lysine/DNA-complexes could be increased by a factor of two to three orders of magnitude, depending on the cell line under consideration. However, the necessity for adding virus particles to the transfection cocktail may raise the risk of gene therapy to unacceptably high levels. PEI is also an interesting substance for application in gene therapy since PEI-containing DNA complexes can be injected without adverse effects into organs such as the brain. Positive, i.e., transfected, cells could be detected a few days later in such experiments.[57]

Cationic Lipids

Most of the numerous commercially available transfection kits use cationic liposomes as transfection agent. The basic unit, Figure 2.4a, of such molecules mimics the natural lipids, which are also found as part of the cell membrane. The formulation and the exact chemistry of these molecules is usually proprietary, hence it is difficult to postulate a mechanism or to suggest improvements of either the molecule or the procedure. The protocols are simplified for routine use and do not present much information. Typical procedures state, for example: add DNA to buffer A, add enhancer and mix with B, wait 5 minutes and add to cell culture, optimize the DNA concentration and the volume of the mix added to cells. In spite of the apparent simplicity, however, it is strongly advised to carry out the short optimization indicated by the protocol in order to obtain the maximum transfection efficiency. While kits are expensive, they require little previous experience and lead quickly to results. However, in our hands at least, these kits were not superior to calcium phosphate or PEI-mediated transfections.

In 1997, the most promising kits at that time available for transfection were compared in our laboratory. All experiments were performed using a standard cell line, human embryo kidney cells[58] (293 HEK), which is routinely used in our group for the rapid production of small amounts of proteins, because it is usually easy to transfect. In this investigation we observed no major differences between the kits. Similar expression levels of the target proteins were reached for most of the methods (Fig. 2.7). The main differences among the kits were in regard to reproducibility and "user-friendliness." At present, most of the involved reagents still need to be stored in frozen form. Some lipids are sensitive to the presence of serum, while others are not. Toxicity is another concern and only a few reagents can be left in contact with the cells for more than a few hours.

Calcium Phosphate

The calcium phosphate technique is probably at present the most widely used transfection method. For calcium phosphate-mediated transfection the DNA is coprecipitated with calcium and phosphate ions. The DNA-binding capacity of the precipitate is high and under optimized conditions (crystal diameter $<< 1 \mu m$) the particles can contain up to 30% of their mass in the form of DNA. For best results, calcium should be used in excess (125 mM), while

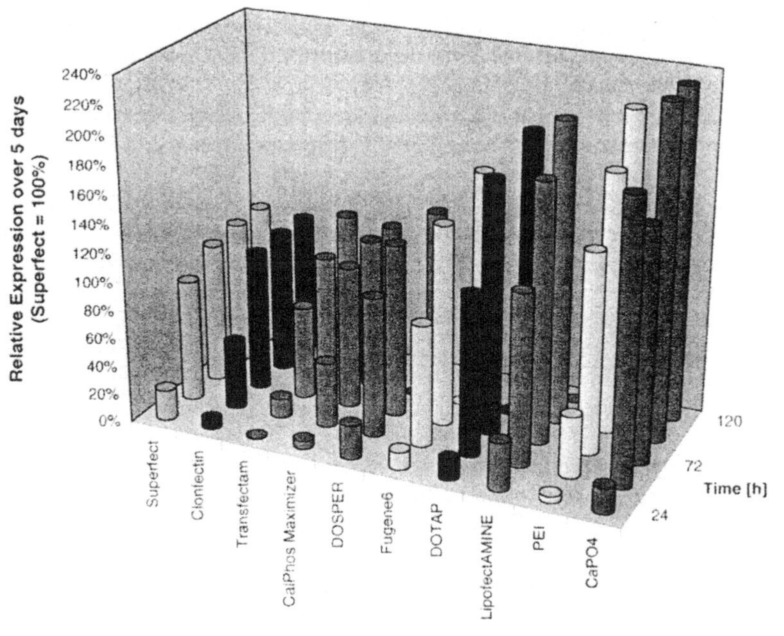

Fig. 2.7. Comparison of commercially available transfection reagents (kits). 293 HEK cells were transfected with one type of plasmid DNA using optimized protocols and expression of recombitant fluorescent protein was measured within the living cells (according to P. Girard, diploma work, Department of Chemistry, Swiss Federal Institute of Technology, 1997).

the phosphate (0.75mM) should be the limiting factor controlling the formation (speed, size) of the precipitate.[59] The precipitate interacts with the cell surface and particles can be taken up by endocytosis. Inside the cell, the crystals dissolve and the DNA is released. Since only calcium and phosphate ions accompany the DNA, no toxic or "unnatural" compounds remain within the cell.

Though several supplier offer commercial kits, many users prefer to prepare their own buffers for calcium phosphate transfections. This adds some time for preparation, sterile filtration and testing. However, if suitably stored, the solutions are stable for years, so large quantities can be prepared at one time. A disadvantage of the method is that the crystal formation is extremely sensitive to differences in pH, which directly affects the solubility. Nevertheless, the method has been used successfully by many groups and has even shown potential for application at large scale (> 10 L).[60]

Conclusions

The perfect transfection method, in which a minute amount of DNA enters the cell by a specific and well-characterized mechanism to be subsequently targeted to the cell's nucleus and the chromosomes within, does not exist; at least as far as man-made transfection agents are considered. Some viral systems, however, do come close to this ideal. One viral particle consisting of protected and packed DNA can infect with a high specificity a particular type of cell in an organism. Such an efficiency and specificity is not achieved by any of the chemical or physical transfection methods outlined in this chapter and the mechanisms of viral infections merit further study. For lack of sufficiently active nonviral systems, viruses have in the past been used for gene therapy. However, in this case some of the possible side-effects (toxicity, infection risk) prevent the method from being ideal.

The perfect transfection agent will be the product of a considerable synthetic effort. A polycationic base polymer needs to be supplied with several functions, including a specific linking group for attachment to the surface of a particular cell (type) and another one for targeting the DNA into the nucleus. The first can be achieved by using one of the indicated receptor sequences. Active import of the DNA into the nucleus may be induced by the peptidic sequence -Pro-Pro-Lys-Lys-Lys-Arg-Lys-Val- , which serves exactly this (targeting) function in nuclear proteins.

Even for the best method, the lysosomes still seem to constitute a considerable barrier. Another problem, which is at least partially linked to endocytosis, is the poor transfectablility of cell lines that lack certain surface molecules or show changes in the endocytotic pathways. A transfection method, which does not depend on endocytosis would be of basic advantage. The ability to circumvent endocytosis as the only way of transporting foreign molecules into cells, would have consequences for other fields as well. Similar strategies could be used, for example, to place toxic drugs directly into (tumor) cells, without affecting healthy neighboring cells.

Definitions

DNAse: (deoxyribonuclease) An enzyme that cuts a strand or completely hydrolyzes DNA into deoxyribonucleotides.

Endocytosis: Uptake of material into a cell by an invagination of the cell membrane and its internalization in a membrane-bounded vesicle (endosome).

Gene: Region of DNA that controls a discrete hereditary characteristic, usually corresponding to a single protein or RNA. This definition includes noncoding regulatory sequences and introns.

Lysosome: Small membrane-bounded organelle in eukaryotic cells containing digestive enzymes, which are typically most active at the acidic pH of 4-5 found in the lumen of lysosomes.

Mitosis: Division of the nucleus of an eukaryotic cell involving condensation of the DNA into visible chromosomes and disruption of the nuclear membrane.

Nucleus: Prominent membrane-bounded organelle in eukaryotic cells containing the genomic DNA organized into chromosomes.

Phagocytosis: Process by which relatively large particles, e.g., bacterial cells, are internalized by certain eukaryotic cells.

Plasmid: Small circular DNA molecule that can replicate independently of the genomic DNA. Used extensively as a vector for DNA cloning. Is usually amplified in *E. coli* from where it can be isolated and used for many applications, e.g., gene transfer into eukaryotic cells.

Transcription: Copying of one strand of DNA into a complementary RNA sequence by the enzyme RNA polymerase.

Transfection: Introduction of foreign DNA into cells in culture, usually followed by the expression of the introduced genes.

Translation: Process by which the sequence of nucleotide in a messenger RNA molecule directs the incorporation of amino acids into protein; occurs on a ribosome.

Structures of Frequently Used Molecules

DOPE[h] (dioleoyl phosphatidylethanolamine)
DOSPA[i] (2,3-dioleyloxy-N-[2(sperminecarboxamido)ethyl]-N,N-dimethyl-1-propanaminium trifluoroacetate)
DOTAP (N-[1-(2,3-dioleoyloxy)propyl]-N,N,N-trimethylammonium methylsulfate)
DOSPER (1,3,di-oleoyloxy-2-[6-carboxy-spermyl]-propylamide)
DOGS[j] (dioctadecylamidoglycylspermine)

References

1. Sambrook J, Fritsch EF, Maniatis T. Molecular Cloning: A Laboratory Manual. 2nd ed, book 1. Coldspring Harbor Laboratory Press, 1989.
2. Goeddel DV, Kleid DG, Bolivard F et al. Expression in *Escherichia coli* of chemically synthesized genes for human insulin. Proc Natl Acad Sci 1979; 76:106-110.
3. Urlaub G, Chasin LA. Isolation of Chinese hamster cell mutants deficient in dihydrofolate reductase activity. Proc Natl Acad Sci 1980; 77:4216-4220.
4. Jordan M, Kohne C, Wurm FM. Calcium-phosphate mediated DNA transfer into HEK-293 cells in suspension: Control of physico-chemical parameters allows transfection in stirred media. Cytotechnology 1998; 26:39-47.
5. Wurm FM, Bernard A. Large-scale transient expression in mammalian cells for recombitant protein production. Curr Opin Biologicals 1999; 10:156-159.
6. Watson J, Crick FH. Molecular structure of nucleic acid. A structure for deoxyribose nucleic acid. Nature 1953; 171:737-738.
7. Mislick DA, Baldschwieler JD. Evidence for the role of proteoglycans in cation-mediated gene transfer. Proc Natl Acad Sci 1996; 93(22):12349-12354.
8. Mounkes LC, Zhong W, Cipres Palacin G et al. Proteoglycans mediate cationic liposome-DNA complex-based gene delivery in vitro and in vivo. J Biol Chem 1998; 273(40):26164-26170.
9. Rosenkranz AA, Yachmenev SV, Jans DA et al. Receptor-mediated endocytosis and nuclear transport of a transfecting DNA construct. Exp Cell Res 1992; 199(2):323-329.
10. Kircheis R, Kickler A, Wallner G et al. Coupling of cell-binding ligands to polyethaleneimine for targeted gene delivery. Gene Ther 1997; 4:409-418.
11. Choi YH, Liu F, Par JS et al. Lactose-poly(ethylene glycol)-grafted poly-L-lysine as hepatome cell-targeted gene carrier. Bioconjug Chem 1998; 9(6):708-718.
12. Zanta MA, Boussif O, Adib A et al. In vitro delivery to hepatocyte with galactosylated polyethyleneimine. Bioconjug Chem 1997; 8(6):839-844.
13. Liang WW, Shi X, Deshpande D, Malanga CJ et al. Oligonucleotide targeting to alveolar macrophages by mannose receptor-mediated endocytosis. Biochim Biophys Acta 1996; 1279(2):227-234.
14. Mislick KA, Baldeshcwieler JD, Kayyem JF et al. Transfection of folate-polylysine DNA complexes: Evidence for lysosomal delivery. Bioconjug Chem 1995; 6(5):512-515.
15. Fujiwara M, Baldeschwieler JD, Grubbs RH. Receptor-mediated endocytosis of poly(acrilic acid)-conjugated liposomes by macrophages. Biochim Biophys Acta 1996; 1278(1):59-67.
16. Uherek C, Fominaya J, Wels W. A modular DNA carrier protein based on the structure of diphtheria toxin mediates target cell-specific gene delivery. J Biol Chem 1998; 273(15):8835-8841.
17. Bailey AL, Cullis PR. Membrane fusion with cationic liposomes: Effect of target membrane lipid composition. Biochem 1997; 36(7):1628-1634.
18. Friend DS, Papahadjopoulos D, Debs RJ. Endocytosis and intracellular processing accompanying transfection mediated by cationic liposomes. Biochim Biophy Acta 1996; 1278:41-50.
19. Zabner J, Fasbender AJ, Moninger T et al. Cellular and molecular barriers to gene transfer ba cationic lipid. J Biol Chem 1995; 270(32):18997-19007.

[h] Neutral helper lipid present in various commercially available transfection reagents.
[i] Lipofect AMINE™
[j] Transfectam™

20. Legendre JY, Szoka FC. Delivery of plasmid DNA into mammalian cell lines using pH-sensitive liposomes: Comparisons with cationic liposomes. Pharm Res 1992; 9(10):1235-1242.
21. Greber UF, Willetts M, Webster P et al. Stepwise dismantling of adenovirus 2 during entry into cells. Cell 1993; 75:447-486.
22. Plotch SJ, O'Hara B, Morin J et al. Inhibition of influenza A virus replication by compounds interfering with fusogenic function of the viral hemagglutin. J Virol; 73(1):140-151.
23. Chavez A, Busquests MA, Pujol M et al. pH-induced destabilization of lipid bilayers by a peptide from the VP3 protein of the capsid of hepatitis A virus. Analyst 1998; 123:2251-2256.
24. Luthman H, Magnusson G. High efficiency polyoma DNA transfection of chloroquine treated cells. Nucleic Acids Res 1983; 11:1295-1308.
25. Wilson SP, Smith LA. Addition of glycerol during DNA exposure enhances calcium phosphate transfection. Anal Biochem 1997; 246:148-150.
26. Zauner W, Kichler A, Schmid W et al. Glycerol enhancement of ligand-polylysine/DNA transfection. Biotechniques 1996; 20:905-913.
27. Kawai S, Nishizawa M. New peocedure for DNA transfection with polycation and dimethylsulfoxide. Mol Cell Biol 1984; 4:1172-1174.
28. El Ouahabi A, Thiry M, Pector V et al. The role of endosome destabilizing activity in the gene transfer process mediated by cationic lipids. FEBS Lett 1997; 414(2):187-192.
29. Cotten M, Wagner E, Zatloukal K et al. High-efficiency receptor-mediated delivery of small and large gene constructs using the endosome-disruption activity of defective or chemically inactivated adenovirus particles. Proc Natl Acad Sci 1992; 89:6094-6098.
30. Kichler A, Mechtler K, Behr JP et al. Influence of membrane-active peptides on lipospermine/ DNA complex mediated gene transfer. Bioconjug Chem 1997; 8:213-221.
31. Boussif O, Lezoualch F, Zanta MA et al. A versatile vector for gene and oligonucleotide transfer into cells in culture and in vivo: Polyethyleneimine. Proc Natl Acad Sci 1995; 92:7297-7301.
32. Graham FM, van der Eb AJ. A new technique for the assay of infectivity of human adenovirus 5 DNA. Virology 1973; 52:456-567.
33. Escriou V, Ciolina C, Helbing LA et al. Cationic lipid-mediated gene transfer: Analysis of cellular uptake and nuclear import of plasmid DNA. Cell Biol Tox 1998; 14:95-104.
34. Dean DA. Import of plasmid DNA into the nucleus is sequence specific. Exp Cell Res 1997; 230:293-302.
35. Balicki D, Beutler E. Histone H2A significantly enhances in vitro DNA transfection. Mol Med 1997; 3:782-787.
36. Vacik J, Dean BS, Zimmer WE et al. Cell-specific nuclear import of plasmid DNA. Gene Therapy 1999; 6:1006-1014.
37. Zhang F, Andreassen P, Fender P et al. A transfecting peptide derived from adenovirus fiber protein. Gene Therapy 1999; 6:171-181.
38. Collas P, Alestrom P. Nuclear localization signals: A driving force for nuclear transport of plasmid DNA in zebrafish. Biochem Cell Biol 1997; 75:633-640.
39. Lechardeur D, Sohn KJ, Haardt M et al. Metabolic instability of plasmid DNA in the cytosol: A potential barrier to gene transfer. Gene Therapy 1999; 6:482-497.
40. Tseng WC, Haselton FR, Giorgio TD. Mitosis enhances transgene expression of plasmid delivered by cationic liposomes. Biochim Biophys Acta 1999; 1445:53-64.
41. Own observation with $CaPO_4$ technique and synchronized cells (unpublished).
42. Wilke M, Fortunati E, van den Broek M et al. Efficacy of a peptide-based gene delivery system depends on mitotic activity. Gene Therapy 1996; 3:1133-1142.
43. Wolff JA, Malone RW, Williams P et al. Direct gene transfer into mouse muscle in vivo. Science 1990; 247:1465-1468.
44. Chu G, Hayakawa H, Gerg P. Electroporation for the efficient transfection of mammalian cells with DNA. Nucleic Acids Res 1987; 15:1311-1325.
45. McCutchan JH, Pagano JS. Enhancement of the infectivity of simian virus 40 deoxyribonucleic acid with diethyl aminoethyl-dextran. J Natl Canc Inst 1968; 41:351-357.
46. Sussman DJ, Milmann G. Short-term, high-efficiency expression of transfected DNA. Mol Cell Biol 1984; 4:1641-1643.

47. Pouton CW, Lucas P, Thomas BJ et al. Polycation-DNA complexes for gene delivery: A comparison of the biopharmaceutical properties of cationic polypeptides and cationic lipids. J Cont Rel 1998; 53:289-299.
48. Nead MA, McCance DJ. Poly-L-ornithine-mediated transfection of human keratonocytes. J Invest Dermatol 1995; 105:668-671.
49. Chemin I, Moradpour D, Wieland S et al. Liver-directed gene transfer. A linear polyethylenimine derivative mediates highly efficient DNA delivery to primary hepatocytes in vitro and in vivo. J Virol Hepat 1998; 5:369-375.
50. Personal communication EJ Schlaeger at F. Hoffmann La Roche, Basel, Switzerland.
51. Diglio CA, Piper CE, Ferrer JF. An improved synctia infectivity assay for the bovine leukemia virus. 1978; 14:502-505.
52. Cornetta K, Anderson WF. Protamine sulfate as an effective alternative to polybrene in retroviral mediated gene transfer: Implication for human gene therapy. J Virol Meth 1989; 23:187-194.
53. Aubin RA, Weinfield M, Taghavi M et al. Highly effective delivery of foreign DNA to adherent cells via polybrene/DMSO-assisted gene transfer. Meth Mol Biol 1997; 62:319-342.
54. Aubin RA, Weinfield M, Mirzayans R et al. Polybrene/DMSO-assisted gene transfer. Generating stable transfectants with nanogram amounts of DNA. Mol Biotechnol 1994; 1:29-48.
55. Kukowska LJF, Bielinska AU, Johnson J et al. Efficient transfer of genetic material into mammalian cells using Starburst polyamidoamine dendrimers. Proc Natl Acad Sci 1996; 93(10):4897-4902.
56. Zenke M, Steinlein P, Wagner E et al. Receptor-mediated endocytosis of transferrin-polycation conjugates: An efficient way to introduce DNA into hematopoietic cells. Proc Natl Acad Sci 1990; 87:3655-3659.
57. Abdallah B, Hassan A, Benoist C et al. A powerful nonviral vector for in vivo gene transfer into the adult mammalian brain Polyethylenimine. Hum Gene Ther 1996; 16:1947-1954.
58. Pear WS, Nolan GP, Scott ML, Baltimore D. Production of high titer helper-free retroviruses by transient transfection. Proc Natl Acad Sci 1993; 90:8392-8396.
59. Jordan M, Schallhorn A, Wurm FM. Transfecting mammalian cells: Optimization of critical parameters affecting calcium-phosphate precipitate formation. Nucleic Acids Res 1996; 24:596-601.
60. Jordan M. Transient gene expression in mammalian cells based on the calcium phosphate transfection method. In: Al-Rubeai M, ed. Cell Engineering. Vol. 2. Kluwer Academic Publications, 2000:56-79.

CHAPTER 3

Affinity Precipitation:

Stimulus-Responsive Polymers for Bioseparation

Ruth Freitag

Introduction

The modern biotechnology industry has provided the medical community with a new type of pharmaceutical, namely recombinant proteins and peptides. The number of such protein-based drugs is already impressive and expected to increase considerably in the future, as the function and gene sequence of more and more proteins is discovered, for example, as a result of the human genome project and related activities (genomics, proteomics). Soon gene therapy may bring about the next revolution in medical treatment, where for the first time it will be possible to correct the genetic causes of a disease rather than to just counteract the consequences of that genetic defect, i.e., the symptoms of the disease. Gene vaccines present another evolving medical possibility, which has been shown to have some significant advantages over the prophylactic measures taken at present against certain infectious diseases. Biopharmaceuticals, such as proteins and nucleic acids, thus extend the possibilities of medical treatment and thereby improve the quality of life. From an engineering point of view, however, the production and especially the purification of these biopharmaceuticals poses a considerable challenge.[1,2]

Proteins, i.e., the majority of the current biopharmaceuticals, are fragile, easily denatured substances. They are generally found in a complex environment such as cell, culture media or lysates (inclusion bodies). The concentration of the target molecule in the raw feed is often low, while extremely high final purities have to be reached in order to fulfill current legal requirements(see Table 3.1 for details). The production scale in the biopharmaceutical industry is often smaller than in the conventional pharmaceutical industry, i.e., in the kg/year rather than the t/year range. Speed (throughput) is another issue since prolonged exposure to a nonsterile environment or certain impurities (proteases) may lead to significant product degradation. Last but not least, the biotech industry, as any other, is governed by economic demands.[3]

Conventional pharmaceuticals, on the other hand, are usually small biochemically active molecules which inhibit or enforce a metabolic function. Such substances can be purified by a host of well-understood separation methods including extraction, precipitation, filtration, refraction and crystallization for highest purity. Most of these separation methods cannot be used for the separation of proteins due to the prevalence of harsh "nonphysiological" conditions such as elevated temperatures, extreme pH values and organic solvents. Thus, the challenge of industrial bioseparation can be summed up as: The need to perform an economically sound, high-resolution separation on a large scale, while maintaining "physiological" conditions throughout.

Synthetic Polymers for Biotechnology and Medicine, edited by Ruth Freitag.

Table 3.1. Requirements for the final purity of a recombinant pharmaceutical intended for parenteral use in humans (combined from information given in reference 1)

Contaminant/Impurity	Maximum Residual Allowed
Host cell DNA	10 pg/dose
Endotoxin	0.25 units/ml
Foreign proteins (total)	< 1 %
Host cell protein	< 10 ng/dose
Protein A	< 10 ppm
Aggregates	< 1 %

The fact that product isolation contributes heavily to the overall production costs is well known in the bioindustry.[4,5] Another important issue is the time-to-market, i.e., the time required for process development and scale up to production. The latter aspect, in particular, seems to argue in favor of the continued use of well-established methods such as chromatography.[6] However, on close scrutiny, both the costs and the process development and adaptation time may in fact be improved by the rapid integration of evolving new separation principles into the category of "established techniques."[7,8]

The goal is a reduction in the number of steps and an increase in the overall efficiency of a given downstream process. New developments in the material sciences will be essential in advancing this issue. Smart or intelligent materials, such as stimulus-responsive polymers capable of sensing a certain change in their environment and reacting in a corresponding and predetermined way, offer many possibilities in this regard.

The Role of Affinity Separations in Product Isolation

The principal problem of many industrial scale preparative bioseparations is the quick and specific isolation of a lowly concentrated target molecule from a complex matrix containing a large number of impurities and contaminants. Various strategies exist to tackle this problem. One popular approach consists in first capturing the entire class of molecules to which the target substance belongs, e.g., all proteins, thus removing the water and rendering subsequent separation steps more efficient. Another strategy, which becomes more popular the lower the product concentration and the more complex the matrix, calls for the use of an "affinity separation."[9,10] Affinity separations are based on some form of "biospecific" interaction; examples include antibody antigen-, receptor hormone-, or enzyme substrate- enzyme inhibitor-interactions. The list does not end there and it is certainly not limited to naturally-occurring substances.

The principle of biospecific interactions is the requirement for a steric fit. Each of the two molecules presents a number of interaction sites in a certain three-dimensional arrangement. "Binding" requires the concomitant realization of several of these (usually noncovalent) interactions. In the case of a "perfect fit," the resulting bond can be quite stable due to the synergism between the various interactions. In this context, high affinity constants and low dissociation rate constants represent strong binding. Since a number of different types of interactions (e.g., hydrophobic, electrostatic, H-bridge formation) are involved in creating the complex, the binding strength can easily be fine-tuned and modified.

Nature has certainly optimized the use of "biospecific interactions," however, she has no monopoly on them. The underlying principle of exploiting steric congruity to distinguish between closely related molecules may also be used by chemists to systematically synthesize materials for "affinity separations". Examples include the molecular imprinted polymers discussed in Chapter 7, but also the use of certain dyes as group-specific ligands or the use of combinatorial approaches (such as phage display of combinatorial chemistry) to identify molecules with a well-defined affinity towards the target molecule.[11-13] For the removal of an impurity, the strongest possible interaction should be used. In the case of product isolation, however, the recovery of the intact product requires controlled dissociation of the affinity complex at some point. In affinity chromatography, currently the most common type of affinity separation, dissociation rate constants between 10^{-4} M and 10^{-8} M are recommended.[9] In the case of much higher values the binding is most likely too weak, while much lower constants will make the recovery of the product difficult.

Affinity separations use the highly specific binding of the target molecule to the affinity mediator (ligand) to force said target molecule into a second phase, which is not easily accessible to the majority of the contaminants/impurities. In affinity chromatography, for example, the affinity ligand is bound to a stationary phase packed into a chromatographic column. The product containing feed is passed through the column, where only the product molecules are retained, while the impurities pass without interaction. The separation is easily achieved and the product can usually be recovered in highly concentrated and purified form. Affinity chromatography has been used successfully in many biotechnical downstream processes, especially in the area of products from mammalian cell culture.[14]

However, as mentioned before, the main allure of using affinity techniques stems from the fact that specific product capture is possible even from very complex and/or diluted matrices. Affinity steps are thus best used early on during a multi-step purification process (also since the unavoidable ligand leakage presents less of a problem in this case). In this regard affinity chromatography has some serious drawbacks. The large-scale potential of the method is limited and columns do show a tendency to clog when typical raw feeds such as cell culture supernatants or bacterial lysates are processed. Fluidized and expanded bed adsorption has recently begun to ameliorate some of these problems.[15]

Many other separation principles, which may in fact be much more suited to the needs of large-scale bioseparation than chromatography, may also be "improved" in terms of selectivity by the introduction of the "affinity motif." Among them are: affinity extraction in aqueous two-phase systems, affinity (ultra-)filtration, but also affinity precipitation, the topic of this chapter. Extraction, filtration and precipitation are unit operations, which are routinely used at large scale in the chemical and pharmaceutical industry. Why is the progress of their affinity variants so slow? Two factors may possibly contribute. One is the fact that higher affinity constants are necessary in these processes than in affinity chromatography.[16,17] In affinity chromatography the target molecules may interact repeatedly with the stationary phase during their progress along the column, since this is a typical multiple interaction process. The above-mentioned alternative methods must be considered single interaction operations. In this case only high affinity constants assure good yields, albeit with the ensuing problems of product release once the separation has been achieved.

Adding further to the problem of binding strength is the fact that in nature the highest affinities and specificities are found for large affinity ligands such as proteins, which have many interaction points. However, proteineous affinity mediators are less popular in downstream processing, since leakage can never be avoided totally and the removal of the affinity ligand from the product may be challenging in such cases. Limited life time and high costs are additional issues for such ligands. Small affinity mediators such as triazine dyes, peptides, and even certain chelated metal ions (Cu(II), Ni(II), Zn(II)), on the other hand, are easily removed from the product, for example, by gel filtration. In addition these ligands are more likely to withstand harsh cleaning and sanitizing conditions.[18] Their only drawback is a possible lower affinity constant.

However, the binding strength between a small affinity mediator and the target molecule can be enhanced in several ways if necessary. One possibility is to use the concept of avidity, i.e., to allow for multiple binding through a high local concentration of the affinity mediator. The other possibility is to systematically design the affinity mediator for high affinity, for example through combinatorial chemistry. After all, the strongest noncovalent binding found in nature takes place between the protein avidin and the small ligand biotin.[19] The design of the affinity mediator is not part of this overview. However, it should be kept in mind that such ligands can be produced and that the affinity principle is therefore applicable to a much wider range of separation technologies than just chromatography.

The Principle and Application of Affinity Precipitation

In affinity precipitation, the affinity reagent is allowed to interact with the target molecule in free solution. Once they have been formed, the affinity complexes and sometimes the surplus reagent molecules are precipitated out of the solution in some way or another. The target molecule is recovered by interruption of the affinity interaction and finally the affinity reagent is recycled if possible. Affinity precipitation has many advantages over conventional bioseparation operations including affinity chromatography. Especially important are certainly:

1. the inherently high concentration and purification factors,
2. the potential for quick and efficient removal of the target molecule from the raw feed,
3. the unsurpassably low influence of mass transfer limitations and steric hindrances during affinity complex formation (and putatively also during complex break-up),
4. the fact that the process is relatively independent of the feed composition (the viscosity may, for instance, vary over a wide range without posing a major handicap),
5. the good scale up potential (theoretically at least, affinity precipitation can be carried out at the hectoliter as well as at the milliliter scale).

In the pertinent literature two principal types of affinity precipitation are distinguished. In the so-called primary effect affinity precipitation (Fig. 3.1) the precipitation is the direct result of the formation of the affinity complex. For this type of affinity precipitation at least two binding sites are required on the target molecule. Then it is possible to cross-link a number of target molecules by bivalent affinity mediators. Once the complex is large enough, it will precipitate. Perhaps the best known example for this type of separation is the immunoprecipitation, where bivalent antibodies are used to precipitate the corresponding multivalent antigen.[20] Immunoprecipitation also gives an idea of some of the disadvantages that may be associated with this type of purification.

1. The application range is restricted to target molecules bearing at least two binding sites for the affinity medication.
2. The concentration of the target molecules in a given feed must be fairly well known beforehand, since both too high and too low a concentration for a given amount of cross-linker will result in the formation of only small associates, which tend to stay in solution.
3. The method is not easily applicable for small target molecule concentrations.
4. Recovery of the target molecules from the complex (and also redissolution of the complexes) may be difficult.
5. The capacity for recycling of the affinity cross-linker is limited.

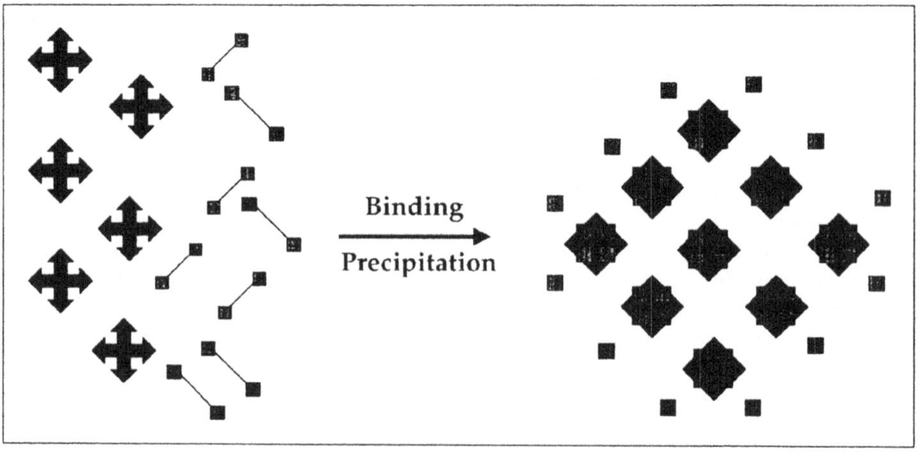

Figure 3.1. Schematic presentation of a primary effect affinity precipitation using a bis-ligand as cross-linker to precipitate a tetrameric protein. (reproduced with permission from reference 59).

Primary effect affinity precipitation is best suited for the purification of proteins showing multiple-point interaction with some of the above mentioned small affinity ligands. Typical target substances for primary effect affinity precipitation are, for example, enzymes, which interact with triazine dyes such as Cibacron blue, but also proteins rich in surface histidine residues, since these protein show strong interaction with immobilized (chelated) transition metal ions such as copper or zinc. The latter effect can be used in a very general manner since the introduction of so-called histidine-tags (i.e., multiple histidine sites) into recombinant proteins is a routine procedure, which normally has little or no influence on the biological activity of the protein.[21,22]

In secondary effect affinity precipitation (Fig. 3.2) affinity binding and enforced precipitation are separate functions of a so-called affinity macroligand (AML). The basis of an AML is a stimulus-responsive ("smart") polymer, which is soluble in water under certain conditions, but becomes insoluble (i.e., precipitates) once a critical parameter has been surpassed (see below for details). In addition the molecule bears an affinity ligand which is responsible for the specific binding of the target molecule. Secondary effect affinity precipitation does not require multivalent target molecules; on the contrary, most applications of this variant are for monovalent targets. Large ligands such as whole antibodies have been used successfully as affinity ligands in secondary affinity precipitation, as has a host of low molecular mass ones. The only caveat in regard to this method concerns the necessity to maintain the affinity interaction even during the precipitation step.[23]

In secondary effect affinity precipitation, which will be the variant considered exclusively from here onward, the redissolution of the affinity complex after precipitation is usually straightforward. The target molecule may be recovered either directly from the precipitate, which is clearly preferable from a process-engineering point of view, or after redissolution of the affinity complex. Authors who investigated the aspect of AML-recycling usually encountered no problems in reusing the affinity reagent several times.[24,25] Below one recently published case study is detailed as a demonstration of the potential of secondary effect affinity precipitation for protein recovery.

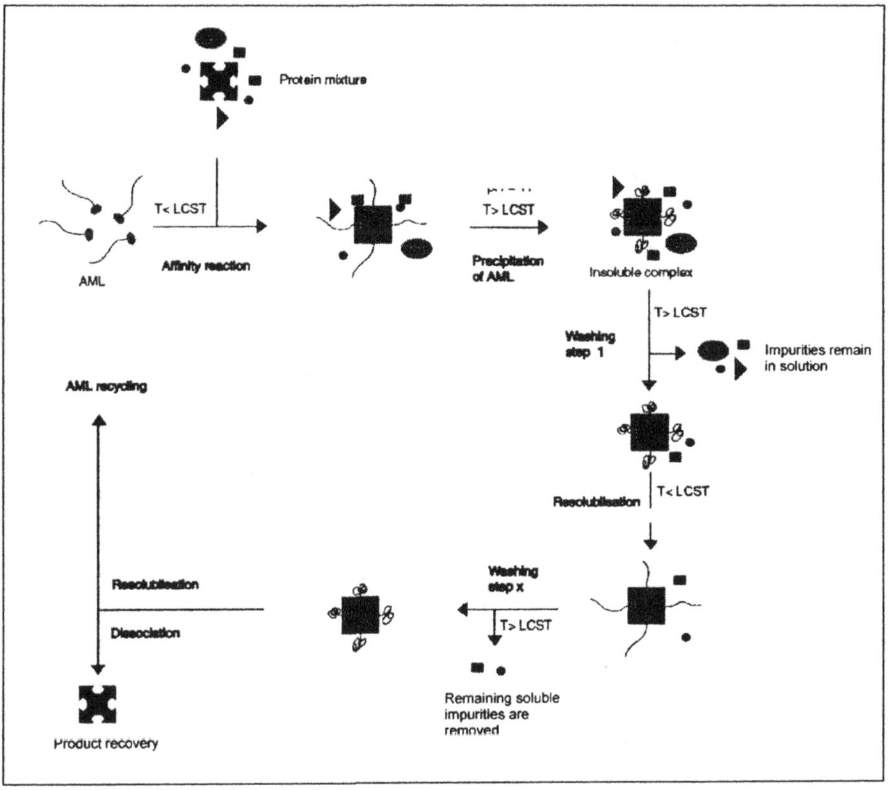

Figure 3.2. Schematic presentation of a secondary effect affinity precipitation using a heterobifunctional affinity macroligand to precipitate the target molecule (reproduced with permission from reference 25).

Recovery of Avidin from Cell Culture Supernatant by Secondary Effect Affinity Precipitation

In this case study[25] a thermoresponsive affinity macroligand was used (see below for stimulus-responsive polymers in general). The base polymer (AML-precursor) was a poly-N-isopropylacrylamide (PNIPAAm) which was produced by chain transfer polymerization (Fig. 3.3). As a result of the particular chain transfer agent employed in this case, each AML-precursor carried a terminal amino group to which putative affinity mediators could be coupled using standard carbodiimid chemistry,[26] a method which generally preserves the ligand's activity to the highest possible degree. For avidin recovery, iminobiotin was used as affinity mediator. The ensuing AML had a high affinity ($K_D \sim 10^{-15}$ M) to avidin and avidin-tagged molecules at elevated pH (ca. 10), but released the avidin readily at lower pH ($K_D \sim 10^{-3}$ M at pH 4). The AML-precursors had an average molecular mass of less than 5000 g/mol, with a polydispersity below 1.2. As a consequence, the solubility behavior of the oligomer was very homogeneous. In pure water the AML showed sharp thermoprecipitation at 31.9° C, Figure 3.4. The critical temperature could be influenced by the addition of salt and other cosolutes, but precipitation remained sharp and reproducible under all circumstances. Quantitative thermoprecipitation occurred within seconds after the critical temperature was surpassed.

For the actual avidin recovery, an AML-preparation ("affinity reagent") was used in which only one out of ten oligomer molecules carried the affinity mediator. This takes into

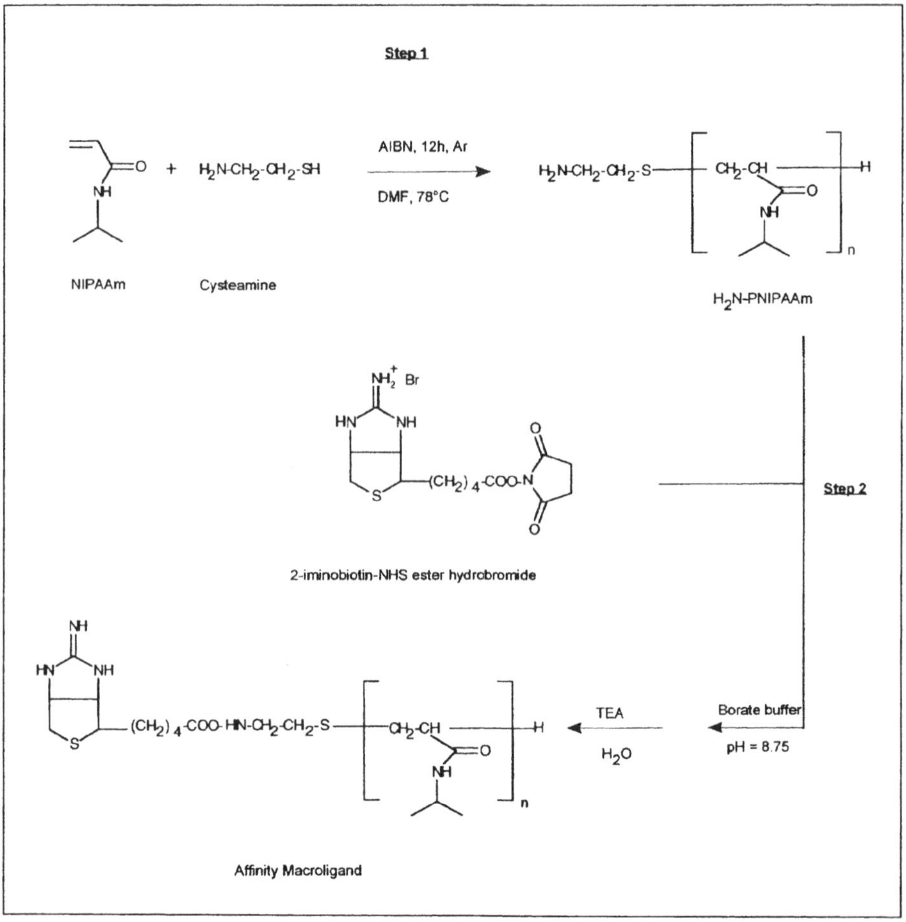

Figure 3.3. Reaction scheme for the synthesis of an oligomeric poly-N-isopropylacrylamide based affinity macroligands through chain transfer polymerization, followed by conjugation of the affinity ligand 2-iminobiotin (reproduced with permission from reference 25).

consideration the fact that the precipitation (flock formation) is actually improved at higher polymer concentration while a much lower amount of the (expensive) conjugated AML may already provide a sufficient product capacity.[27] The avidin capacity of the affinity reagent was 0.16 mg of avidin per mg of reagent. The binding between the AML and the protein (avidin) was improved by the addition of 0.5 M NaCl to the binding buffer (Table 3.2). The avidin concentration, on the other hand, had no influence on the avidin recovery yield, provided that the AML's binding capacity was high enough (no saturation).

In affinity precipitation the equivalent to a washing step is the redissolution and reprecipitation of the AML target protein-complex without breaking the affinity interaction (e.g., through repeated thermocycling in subsequent fresh batches of binding buffer). Up to seven of these cycles were performed with the iminobiotin-activated affinity reagent before the avidin was finally released from the complex by a transfer to the dissociation buffer. As in any washing step, some product is lost during each cycle and the number of cycles must hence represent a carefully optimized compromise between purity and recovery yield. However, the

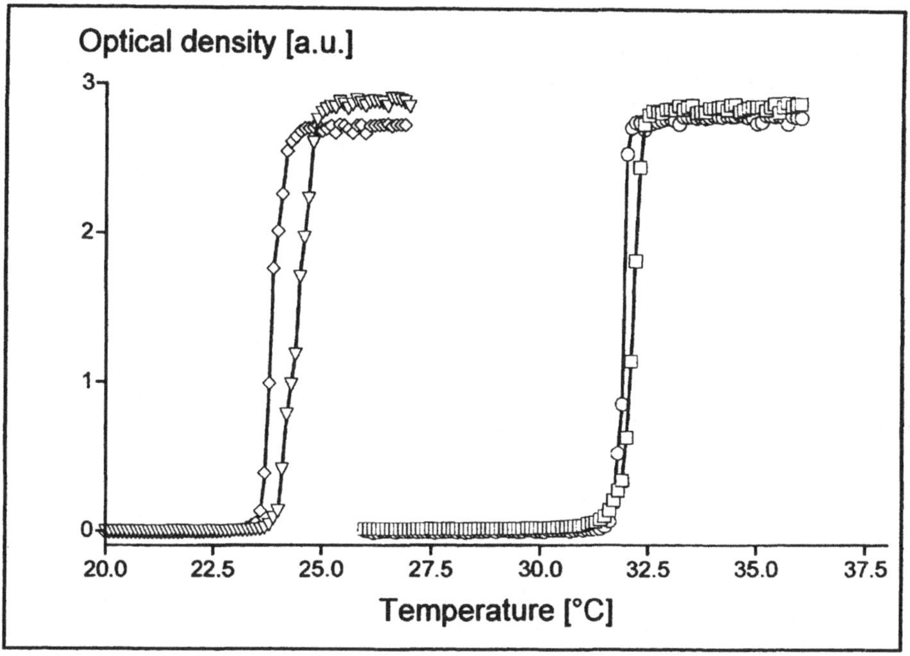

Figure 3.4. Thermoprecipitation curves of a) □ the nonconjugated AML-precursor and ○ the AML in pure water, b) ∇ only the AML in binding buffer,c) ◊ the affinity complex (AMLproduct)(reproduced with permission from reference 25).

loss was minimal in the case of the avidin complex and almost 90% of the product could still be recovered after the seven washing cycles (Fig. 3.5).

The affinity reagent was used to purify avidin from cell culture supernatants containing 5% fetal calf serum; 50 mg of the affinity reagent were added to a cell culture supernatant containing 0.25 mg/ml of avidin. After complete dissolution of the AML, the pH of the solution was adjusted to 10.8 with NaOH. Concomitantly, the NaCl concentration was brought to 0.5 M. Precipitation was induced by raising the temperature 2° C above the critical temperature and the precipitate was recovered by centrifugation. The SDS page gel of the different supernatants obtained taken during the purification process is shown in Figure 3.6. The initial sample, S1, shows five major protein bands corresponding to masses between $5x10^4$ and $2x10^5$ g/mol in addition to the avidin band at $1.6x10^4$ g/mol. After the first induced precipitation, the majority of protein contaminants remained in the supernatant S2; however, two more washing steps were necessary (S3, S4) to remove all contaminants. Avidin was subsequently released from the AML in the dissociation buffer and recovered in supernatant S5. Only 10% of the avidin were lost during the cleaning step, and the total recovery yield of avidin was 90%. The residual protein contaminants of the purified avidin was below the detection limit, and the specific content of avidin in the total protein content was enhanced about 14-fold compared to the initial cell culture supernatant. In a control experiment, the nonconjugated AML-precursor rather than the AML was used to attempt "affinity precipitation" of avidin from the supernatant under otherwise identical condition. No protein could be found in the dissociation buffer supernatants in this case. This clearly indicates that there is no (detectable) nonspecific coprecipitation of either avidin or the cell culture supernatant components by the nonconjugated AML precursor.

Table 3.2. Influence of salt addition, avidin concentration, pH and repeated washing cycles on avidin recovery by affinity precipitation (reproduced with permission from reference 25)

Product Concentration	Buffer / Feed	NaCl Concentration	Recovery Yield
0.49 mg/ml	0.1 M carbonate, pH 10.8	none	75%
0.49 mg/ml	0.1 M carbonate, pH 10.8	0.5 M	87%
1.00 mg/ml	0.1 M carbonate, pH 10.8	0.5 M	85%
0.25 mg/ml	0.1 M carbonate, pH 10.8	0.5 M	90%
0.49 mg/ml	0.1 M carbonate, pH 9.8	none	33%
0.49 mg/ml	0.1 M phosphate, pH 8.0	none	7%
0.25 mg/ml	0.1 M carbonate, pH 10.8 / 5 mg/ml lysozyme	0.5 M	93%*
0.25 mg/ml	Cell culture medium (5% FCS)	Medium composition proprietary	90%*

*residual contamination by other proteins below the detection limit

Smart Polymers for Affinity Precipitation

Reversibly water-soluble macromolecules can be responsive to a number of external stimuli. The most commonly used parameters are:

1. the temperature
2. the ionic strength
3. the pH

Light irradiation, as well as the addition of cross-linking agents, surfactants, or oppositely charged polyelectrolytes to a charged AML preparation are less commonly used but also valid approaches to enforce precipitation in certain cases.

In order to be water-soluble in the first place, the macromolecule needs to contain some "hydrophilic" groups, such as charged or H-bridge forming units. If these groups' ability of mediating a contact between the macromolecule and the water molecules is reduced, the macromolecule will no longer stay in solution, but will form (usually hydrophobic) intra- or intermolecular contacts instead. The efficiency of H-bridge formation decreases with increasing temperature. A critical temperature may hence be observed for such molecules. Charged polymers may be simply precipitated by reducing or neutralizing the charge density, e.g., through the addition of oppositely charged (poly-)ions. In cases where the charge density depends on the pH, as for some poly-(meth-)acrylic acids, a pH shift may bring about precipitation. Below the various types of possible AML-precursors are discussed in detail.

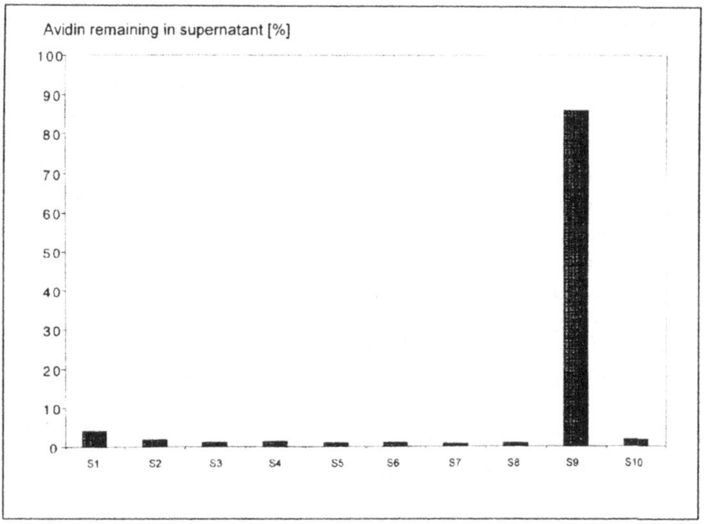

Figure 3.5. Effect of repeated "washing" of the affinity complex prior to product release on product recovery. After the initial formation and precipitation of the affinity complex from the product containing feed (S1), the complex was repeatedly thermoprecipitated and redissolved in fresh binding buffer (S2-S8). Afterwards the bound avidin was released in dissociation buffer (S9,S10).

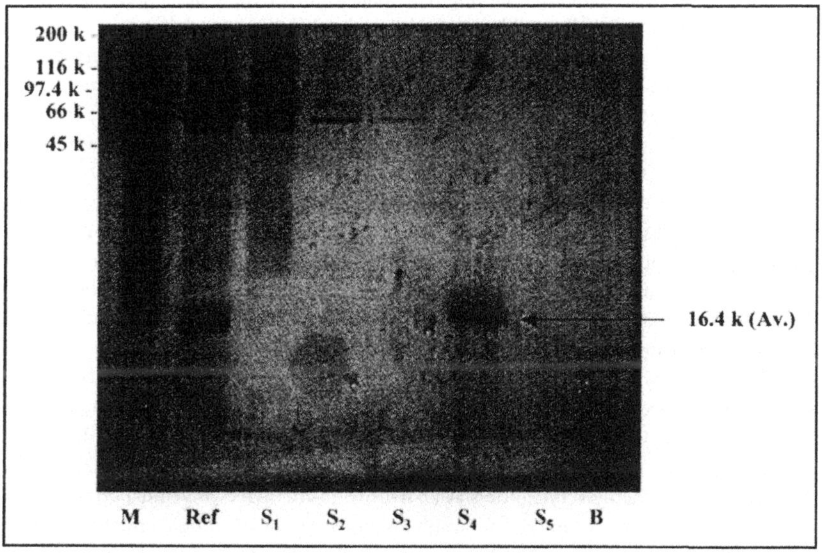

Figure 3.6. SDS-PAGE gel of samples taken during affinity precipitation of avidin from cell culture supernatant. M: molecular mass markers, Ref: avidin containing feed, S1: feed after affinity precipitation, S2 and S3: supernatants obtained during "washing" cycles, S4: first dissociating buffer supernatant (release of avidin), S5: second precipitation of the AML from dissociating buffer, B: buffer samples (reproduced with permission from reference 25).

Thermosensitive AML

The solubility of certain uncharged reversibly water-soluble polymers is known to depend strongly on the temperature, and usually a lower critical solution temperature (LCST) is observed above which the polymer is no longer soluble.[28] At low temperature, the monomeric units of the polymer interact preferentially with the water molecules, and the polymer is dissolved in the form of a loose coil. As the temperature increases the efficiency of the H-bridges in stabilizing the polymer water contacts decreases. At the same time, the entropy driven hydrophobic interactions between the monomeric units become more powerful.[29] At the critical temperature the balance is shifted and the dense "globule"-state of the polymer, dominated by direct interactions between the monomeric units themselves, becomes more stable.[30] The next step is usually the rapid aggregation of the polymer globules into large, precipitating aggregates. The consistency of the aggregates can be influenced to some extent by the process parameters.[23,31]

Thermosensitive polymers (oligomers) are perhaps the most promising AML candidates, since:

1. Such molecules are rather inert and less prone to nonspecific interaction with contaminating substances than, for example, charged AML.[32]
2. Thermoprecipitation calls only for heating/cooling of the system, no chemicals need to be added.
3. The critical temperature can be adjusted over a wide temperature range and thermoinduced affinity precipitation has been described between 4°C and 70°C.[33] Even heat sensitive products can thus be recovered.
4. It is possible to synthesize a thermoresponsive AML in a way that a conjugation site is added independently of the monomeric units responsible for thermoprecipitation, e.g., at either terminus). Standard coupling reactions, as for example developed for affinity chromatography, can equally well be used with such AML-precursors.[34]

A large number of thermosensitive polymers have been described in the pertinent literature, usually produced by chemical synthesis. The can be divided into three categories:

1. Homopolymers made from one type of monomeric unit: In such molecules the individual unit strikes a careful balance between hydrophobicity and hydrophilicity. At present, a prediction of the solubility behavior from just the chemical structure is not possible.[35] On the other hand, small differences in structure may have a considerable influence on the LCST. It has been observed that all other factors being equal, acrylamides have a lower LCST than methacrylamides and polymer with aliphatic carbon side chains in the monomeric units have a lower LCST than those with cyclic ones.[28] The tacticity of a molecule is also of influence and may change the LCST by several °C, while the average molecular mass apparently has little effect.[36]
2. Copolymers with a statistical distribution of two or more types of monomeric units: This is perhaps the easiest way to create a thermosensitive polymer, since a fully water-soluble base polymer can usually be transformed into a thermosensitive one by the introduction of hydrophobic units. An example is the recently published case, where the LCST of a set of poly-(N,N-dimethylacrylamide)stat-(7-hydroxy-3-methyl-4-vinyl indanon copolymers was varied between 10° C and 85° C simply by adjusting the indanon content of the molecules between 6% and 15%.[37] While this approach is attractive, it also has some drawbacks. The resulting copolymers will vary widely in comonomer content and distribution. The result is often a broad LCST range and a nonuniform and somewhat unpredictable precipitation behavior.[38]
3. Block-copolymers: These polymers consist of consecutive blocks of monomeric units of a given type (e.g., AAAAAAAABBBBBBBCCCCCCC.....). In many ways block copolymers combine the advantages of the homopolymers and the statistical copolymers. Each of

the blocks shows a uniform, predictable behavior which is often not much influenced by the other blocks.[39] For example, in a statistical copolymer, the LCST (and many other physical parameters) varies with the comonomer content (and distribution). The copolymer has thus unique features, which distinguish it from the corresponding homopolymers and which often are impossible to predict beforehand. In a block copolymer, the units act independently from each other in the manner of the corresponding homopolymer. If, for instance, the LCST of block A in is surpassed, this part of the molecule will collapse and aggregate and often take the rest of the polymer with it. The synthesis of a block polymer is therefore an interesting way of creating a smart polymer since one part of the molecule could be thermosensitive and another pH-sensitive. However, the challenge of carrying out the actual chemical synthesis may be considerable.

pH- and Ionic Strength-Sensitive Polymers

Polyelectrolytes form the second large class of stimulus-responsive polymers used in affinity precipitation. Some reversibly water-soluble, charged polymers occur in nature (chitosan, alginate, certain cellulose derivatives), others, including many alkyl-(meth-)acrylates, have been synthesized. Polyelectrolytes, such as Eudragit[40] (see below) have been used not only in true affinity precipitations but also for the less specific precipitation of certain proteins through electrostatic interactions.[41,42]

In the case of charged polymers, a number of possibilities exist for reducing the solubilizing effect of the charges and possibly forcing the molecules to precipitate.

1. Polymers like chitosan and some cellulose derivatives carry both ionogenic and hydrophobic groups within the molecule. The ionogenic groups may be protonated or not depending on the pH. If these groups are uncharged, the hydrophobic character of the polymer dominates and the water-solubility is low. If the ionogenic groups are charged, the polyelectrolytic character becomes predominant and the polymer is well soluble in water. The solubility of such molecules is therefore a function of the pH. One example for such a pH-sensitive molecule is Eudragit, at present perhaps the most popular AML-precursor. Eudragit (for example Eudragit S 100 and Eudragit L) is the trade name (Röhm Pharma) of a type of methacrylic acid methyl methacrylate copolymer, which is soluble above a pH of 6 and completely insoluble at a pH of 4.5 in the case of Eudragit S100 (3.5 in case of Eudragit L). The polymer is nontoxic and commonly used in the pharmaceutical industry. In the early applications a pH shift below 4.5 was used to bring about precipitation. More recently, however, it was found that Eudragit can be precipitated even at neutral pH and above by raising the temperature and/or adding calcium ions to the solution (typically 50 mM $CaCl_2$ and 40°C are used).[43] The addition of 1 M NaCl may reduce the nonspecific (ionic) interaction between Eudragit and proteins in general.[44] Affinity mediators can easily be linked to the carboxylic acid groups of the Eudragit by carbodiimid coupling.
2. Polyelectrolytes may also be precipitated through cross-linking by multivalent counterions. An example is the precipitation of alginate by the addition of calcium ions or other di- and trivalent metal ions.[45]
3. The addition of a second polyion of the opposite charge leads to the formation of polyanion/polycation-complexes, which are less soluble in water and usually precipitate. At least one example has been published in which a Cibacron blue conjugated polycation (polyethyleneimine) was used to capture lactate dehydrogenase, followed by precipitation through complex formation with polyacrylic acid.[46]
4. Precipitation through ion pairing with a surfactant of the opposite charge may also be a possibility.

As already indicated in some of the above-mentioned cases, the solubility of a given polymer may be influenced by more than one parameter. The addition of a salt may, for example, reduce the temperature or the pH for which precipitation occurs. A change in temperature will almost always influence the precipitation of a charged polymer to some extent.

Oligomeric AML-Precursors

Conventional affinity macroligands are just what their name suggests, i.e., fairly large molecules, which carry an affinity domain. The reasons for this are more or less historical. The first stimulus-responsive molecules to be used for affinity precipitation were naturally occurring polymers and the early synthetic ones were prepared by standard radical polymerization, which also results in rather large structures. For successful affinity precipitation it is necessary that the AML and not the captured target molecule determines the solubility of the affinity complex. A priori, this should be easier when the AML has a certain size. Conventional AML tend to be rather heterogeneous in size and structure. The synthetic ones are somewhat superior in this regard, however, polydispersities of more than 10 are by no means unusual in the case of radical polymers. This makes the characterization of the AML-precursor rather difficult and may in fact present a major handicap when affinity precipitation is presented to regulatory authorities as part of the production process of a recombinant pharmaceutical. The situation becomes worse if copolymers are necessary for some reason (e.g., for the introduction of the affinity mediator). In such cases the solubility behavior becomes erratic and affinity precipitations using such agents have to be developed for each specific case in particular. Much of the principal flexibility of the method is then lost.

Recently, however, methods for the synthesis of smaller, much more homogeneous synthetic "polymers" (or rather oligomers) have been developed, for example group and chain transfer polymerization. These methods have been shown to be interesting approaches to the synthesis of small but efficient thermosensitive AML-precursors.[25,47] Some pH-sensitive polymers have also been produced by these methods, albeit with other applications than affinity precipitation in mind.[48] The solubility behavior of the oligomeric AML-precursors is very similar to that observed for their larger homologues.[36] Due to the synthesis mechanism, both group and chain transfer oligomers carry a reactive end group, such as an amino, alcohol or carboxylic acid group. It is possible to link the affinity mediator molecule to this end group, thereby preparing AML with a single, well-defined affinity domain. Since the size of the AML is small, high ligand densities in terms of number of ligands per gram of AML are reached in such cases. The next question is, whether such AML, especially the ones carrying large affinity domains (antibodies), still show thermoprecipitation behavior themselves, and if so, whether or not affinity precipitation of a protein product could still be achieved with them.

In the case of conventional AML, the introduction of the affinity mediator usually influences the solubility heavily (see below for details).[49] However, no such influence is usually observed in the case of oligomeric AML with terminal affinity domains, and that not even when the affinity domain (end group) is large or charged.[17,50] It has been postulated that in the case of a terminal affinity domain, the two parts of the AML, i.e., the polymer backbone and the affinity ligand, act independently from each other. Therefore an influence of the affinity ligand on a precipitation temperature would only be expected if the affinity ligand itself had an even lower critical temperature than the polymer backbone. Salts and other cosolutes do exert the usual influence on the precipitation temperature.[51] However, in all investigated cases, both the precipitation and the redissolution remained fast and quantitative. Phase separation and its reversal occurred within a fraction of a degree centigrade and no broad precipitation intervals were ever observed in the case of oligomer-based AML-precursors, as this is sometimes the case for conventional polymeric (copolymeric) AML.

Oligomer-based AML can be used (and may even have advantages) for affinity precipitation of proteins and other large target molecules. The reason why such comparatively small molecules can be successful AML was elucidated by a recent investigation of the thermodynamics of their thermoprecipitation.[51] Apparently, the size of the individual oligomer is of little influence since pronounced association takes place long before the phase transition. In fact, the size of the cooperative unit during phase transition seems to be similar for the small oligomers unlike the case of larger molecules of the same base chemistry.

Introduction of the Affinity Mediator

The creation of the AML requires the introduction of an affinity mediator into the base polymer (AML-precursor). Various approaches to this important step have been described in the literature. The procedure chosen in a given case depends on a number of parameters, some of which are discussed in more detail below.

1. Introduction via a terminal group: In this approach the affinity mediator is linked to an active group at either end of the base polymer. Such groups (including amino, carboxylic acid, alcohol, and epoxy groups) can be introduced into the precursor during the starting of terminating step of the polymerization reaction (e.g., of the group or chain transfer kind). In order to keep the ligand density (capacity) sufficiently high in such cases, introduction via a terminal group should only be attempted for oligomeric AML-precursors. In this case, the technique has certain advantages. Positioned at either end of the AML-precursor, the affinity mediator has almost no influence on the solubility of the base polymer. The affinity remains high, usually equal to that observed for the free affinity mediator, and the affinity constant is similar for all AML of a given type.[17] Since high affinity constants are of utmost importance in affinity precipitation (even more so than in affinity chromatography), AML with terminal affinity domains are preferable over AML with a statistical distribution of affinity ligands. Although it has never been reported, oligomeric AML with terminal affinity domain may be prone to the effect of "stacking" (i.e., self-association), which would lower the separation efficiency considerably. Stacking has been reported for certain cross-linking bis-ligands for primary effect affinity precipitation, for example in the case of Cibacron blue-activated molecules.[52] If stacking can be avoided, introduction of the affniity domain via end group chemistry is characterized by extreme flexibility and yields very homogeneous AML.
2. Introduction via polymer analogous reaction: This is perhaps the most general way of introducing the affinity mediator. It can, for example, also be used for the modification of naturally occurring stimulus-responsive polymers. For this approach, suitable groups in the polymer are activated and subsequently used to couple a number of affinity ligand molecules. Since the affinity mediator is introduced along the polymer backbone, the solubility of the AML tends to differ considerably from that of the base polymer.[44] The approach is statistical and a certain heterogeneity is introduced. The linkage of small ligands such as a triazine dye is usually more straightforward than that of larger proteineous ligands such as antibodies. The affinity constants within an AML preparation may vary in dependency on the position along the polymer backbone where the ligand was introduced. The accessibility of the majority of the ligands will be restricted by the polymer coil and usually the average affinity constant measured for such AML are more than one order of magnitude below that of the free ligand.[23,53] However, since several affinity ligands are linked to the polymer backbone (increase of ligand density), avidity effects due to the possibility of forming multiple interactions between the AML and the target molecule are possible and help to counteract the effect of steric hindrance to some extent.[23,49]The number and spacing of the ligands along the polymer backbone, but also the overall AML concentration, may be

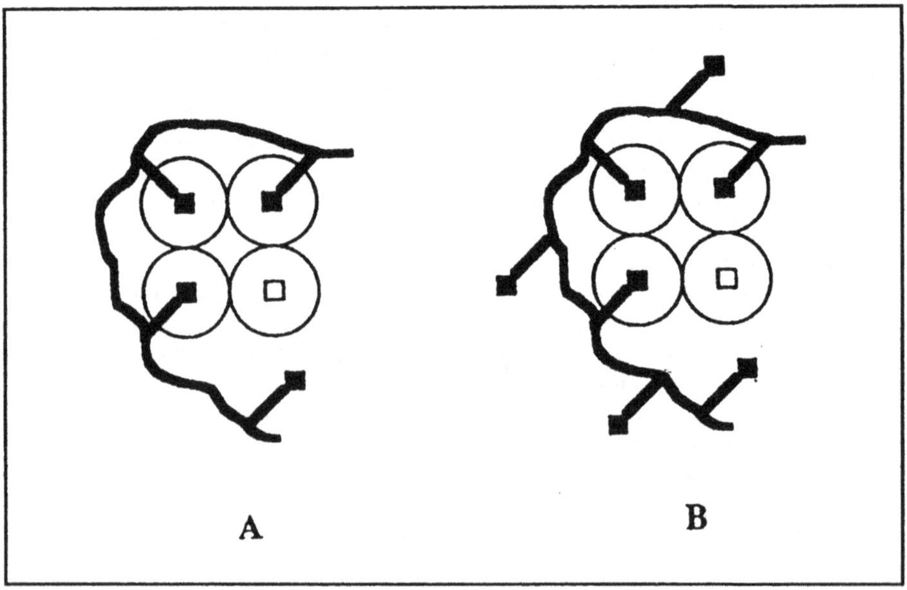

Figure 3.7. Schematic presentation of the "wrapping" effect (A) versus larger associate formation (B) (reproduced with permission from reference 54).

crucial. In some cases so-called "wrapping" has been reported, where the AML enfolds a single target molecule. This yields a soluble one-to-one complex, which shows little tendency to associate with other affinity complexes (Fig. 3.7).[54] In such cases no precipitation may occur.

3. Introduction via copolymerization: This method has some similarities to the one outlined before. Here the AML is produced through a copolymerization of a monomer responsible for the solubility of the final AML and a second one, which carries the affinity mediator or a precursor thereof. The copolymer based AML are also characterized by a pronounced heterogeneity of the ligand density, the affinity constant and the solution behavior.

In all cases the selective "elution" of the target molecule from the AML after successful separation by precipitation may pose an additional problem. Treating the precipitate directly (leaching) would be the easiest approach, but conditions used to release the target molecule from the affinity mediator (pH shift, salt addition, competing agent, etc.) may also resolubilize the AML, especially if a pH- rather than a thermosensitive one is used. Careful design of the polymer backbone and the affinity domain may help to reduce these problems.

Conclusions and Outlook

The appeal of the affinity precipitation approach for bioseparation stems from the concept of combining an well-understood and easily scalable technology like precipitation with the specificity of an affinity interaction. A number of case studies can be found in the pertinent literature, where affinity precipitation (of either type) is used for the isolation or enrichment of biologicals. It should be noted, however, that at present, not a single application of affinity precipitation in an industrial environment has been reported. The reluctance to use a "novel" technology is often high in the pharmaceutical industry due to the increase in regulatory paper work. However, in the case of affinity precipitation, another aspect may be even more important. Until recently, the equipment and the reagents (including the AML) required for affinity precipitation were not commercially available and a do-it-yourself-approach is usually not a good basis for a GMP-type process. However, this has now changed, since at least one company, polyTag Technology, Switzerland, now provides dedicated AML, as well as technical support and services for affinity precipitation.

If one envisions the possible integration of affinity precipitation in a typical multi-stage biotechnical downstream process, the operation will most likely be positioned during the early stages where the advantages such as quick and specific product enrichment come to full play. An interesting combination has been proposed in the combination of affinity precipitation and aqueous two phase system extraction, which could be used even in the case of solid-containing feeds. That was always a problem in pure affinity precipitation since the solids would be co-isolated during the recovery of the precipitate.[55-57] Aqueous two-phase systems form spontaneously when two polymers or a polymer and a salt are dissolved in water above a certain critical concentration.[58] Polyethylene glycol/dextrane and polyethylene glycol/salt systems are the most commonly used ones. Proteins, nucleic acids, cells and cell debris normally partition into the dextran (or salt) phase. Some polymers, including Eudragit, show a tendency to partition into the polyethylene glycol phase instead. If instead of pure Eudragit, an Eudragit-based AML is used, the partitioning behavior in general is almost identical, but the AML will drag the target molecule into the polyethyleneglycol phase. From this phase the target molecule (and the few impurities that will copartition) can easily be recovered by pH-induced affinity precipitation.

The intent of this overview was to present affinity precipitation as a method, which is robust and flexible, but also easy to apply and to vary in scale. Much more work is obviously necessary before affinity precipitation can become a standard bioseparation technology. At present, neither the affinity mediator nor the polymer can be designed to fit a given isolation problem. This adds to the difficulty of introducing this interesting method in a real process environment. Given the importance of the AML-precursor, progress and increased application of affinity precipitation will depend directly on the advances made in the area of "smart polymers".

References

1. Sofer G, Hagel L. Handbook of Process Chromatography. A Guide to Optimization, Scale-Up and Validation. San Diego: Academic Press, 1997.
2. Prazeres DMF, Ferreira GNM, Monteiro GA et al. Large scale production of pharmaceutical-grade plasmid DNA for gene therapy: Problems and bottlenecks. Tibtech 1999; 17:169-174.
3. Wheelwright SM. Economic and cost factors of bioprocess engineering. In: Avis KE, Wu VL, eds. Biotechnology and Biopharmaceutical Manufacturing, Processing, and Preservation. Buffalo Grove: Interpharm Press, Inc., 1996:333-354.
4. De Palma A. Downstream processing operations hold the key to bioprocess cost control. Gen Eng News 1996; 19:19-20.
5. Sadana A, Beelaram AM. Efficiency and economics of bioseparation: Some case studies. Bioseparation 1994; 4:221-235.
6. Sofer GK. Current applications of chromatography in biotechnology. Biotechnology 1986; 4:712-715.
7. Roman M, Brown PR. Separation techniques for biotechnology in the 1990's. J Chromatogr 1992; 592:3-12.
8. Gupta MN, Mattiasson B. Novel technologies in downstream processing. Chem Ind 1994; 5:673-675.
9. Labrou N, Clonis YD. The affinity technology in downstream processing. J Biotechnol 1994; 36:95-119.
10. Luong JHT, Nguyen AL. Novel separations based on affinity interactions. In: Fiechter A, ed. Advances in Biochemical Engineering/Biotechnology. Vol. 47. Berlin, Heidelberg: Springer Verlag, 1992:137-188.
11. Haupt K, Mosbach K. Plastic antibodies: Developments and applications. Tibtech 1998; 16:468-475.

12. Boyer PM, Hsu JT. Protein purification by dye-ligand chromatography. In: Fiechter A, ed. Advances in Biochemical Engineering/Biotechnology. Vol. 49. Berlin, Heidelberg: Springer Verlag, 1992:1-44.
13. Patwardhan AV, Goud GN, Koepsel RR et al. Selection of optimum affinity tags from a phage-displayed peptide library. Application to immobilized copper(II) affinity chromatography. J Chromatogr 1997; 787:91-100.
14. Dyr JE, Suttnar JJ. Separation used for purification of recombinant proteins. J Chromatogr 1997; 699:383-401.
15. Anspach FB, Curbelo D, Hartmann R et al. Expanded-bed chromatography in primary protein purification. J Chromatogr 1999; 865:129-144.
16. Mattiasson B, Kaul R. Affinity precipitation. In: Ngo TT ed. Molecular Interactions in Bioseparations. New York: Plenum Press, 1993:469-477.
17. Eggert M. Baltes T, Garret-Flaudy F et al. Affinity precipitation—An alternative to fluidized bed adsorption. J Chromatogr 1998; 827:269-280.
18. Ramirez-Vick JE, Garcia AA. Recent developments in the use of group specific ligands for affinity bioseparations. Separation and purification methods 1996-97; 25(2):85-129.
19. Green MN. Avidin. Advan Protein Chem 1975; 29:85-113.
20. Freitag R, Scheper T, Schügerl K. Development of a turbidimetric immunoassay for on-line monitoring of proteins in cultivation processes. Enzyme Microb Technol 1991; 13:969-975.
21. Arnold FH. Metal-affinity separations: A new dimension in protein processing. Bio/Technol 1991; 9:151-156.
22. Sassenfeld HM. Engineering proteins for purification. Tibtech 1990; 8:88-93.
23. Galaev IY, Mattiasson B. Affinity thermoprecipitation: Contribution of the efficiency of ligand-protein interaction and access of the ligand. Biotechnol Bioeng 1993; 41:1101-1106.
24. Guoqiang D, Benhura MAN, Kaul R et al. Affinity thermoprecipitation of yeast alcohol dehydrogenase through metal ion-promoted binding with Eudragit-bound cibacron blue 3GA. Biotechnol Prog 1995; 11:187-193.
25. Garret-Flaudy F, Freitag R. Use of the avidin (imino)biotin system as a general approach to affinity precipitation. Biotech Bioeng. In press.
26. Brinkley M. A brief survey of methods for preparing protein conjugated with dyes, haptens , and cross-linking reagents. Bioconjugate Chem 1992; 3:2-13.
27. Senstad C, Mattiasson B. Affinity-precipitation using chitosan as ligand carrier. Biotechnol Bioeng 1989; 33:216-220.
28. Galaev IY, Mattiasson B. Thermoreactive water-soluble polymers, nonionic surfactants, and hydrogels as reagents in biotechnology. Enzyme Microb Technol 1993; 15:354-366.
29. Schild HG, Tirrell DA. Microcalorimetric detection of lower critical solution temperatures in aqueous polymer solutions. J Phys Chem 1990; 94:4352-4356.
30. Wu C. A comparison between the "coil-to-globule" transition of linear chains and the "volume phase transition" of spherical microgels. Polymer 1998; 19:4609-4619.
31. Senstad C, Mattiasson B. Dissertation 1988, Lund University, Sweden.
32. Gupta MN, Mattiasson B. Affinity precipitation. In: Mukherejea RN, ed. Downstream Processing in Biotechnology. India: Tata McGraw-Hill, 1992:43-53.
33. Hoshino K, Taniguchi M, Kitao T et al. Preparation of a new thermo-responsive adsorbent with maltose as a ligand and its application to affinity precipitation. Biotechnol Bioeng 1998; 60(5):568-579.
34. Gubta MN, Mattiasson B. Affinity precipitation. In: Street G, ed. Highly Selective Separations in Biotechnology. London: Chapman and Hall, 1994:7-33.
35. Wang F, Saeki S, Yamaguchi T. Absolute prediction of upper and lower critical solution temperatures in polymer/solvent systems based on corresponding state theory. Polymer 1999; 40:2779-2785.
36. Freitag R, Baltus T, Eggert M. A comparison of thermoreactive water-soluble Poly-N,N-diethylacrylamide prepared by anionic and by group transfer polymerization. J Poly Sci 1994; 32:3019-3030.
37. Freitag R., Vogt S, Mödler M. Thermoreactive displacers for amnon exchange and hydroxyapatite displacement chromatography. Biotechol Progr 1999; 15(3):573-576.

38. Eggert. M. Affinitätspräzipitation als Verfahren zur Proteinaufarbeitung. Entwicklung thermosensitiver Makroliganden für Subtilisin Carlsberg. Dissertation, 1993.
39. Bromberg L, Levin G. Poly(amino acid)-b-Poly(N,N-diethylacrylamide)-b-Poly(amino acid) conjugates of well-defined structure. Bioconjugate Chem 1998; 9:40-49.
40. Kaul R, Guoqiang D, Lali A et al. Affinity isolation of proteins based on precipitation of Eudragit-ligand-protein complexes. In; Pyle DL, ed. Separations for Biotechnology 3. Cambridge: The Royal Society of Chemistry, 1994:242-248.
41. Gupta MN, Guoqiang D, Kaul R et al. Purification of xylanase from trichoderma viride by precipitation with an anionic polymer Eudragit S 100. Biotechnol Tech 1994; 8(2):117-122.
42. Guoqiang D, Kaul R, Mattiasson B. Purification of lactobacillus bulgaricus D-lactate dehydrogenase by precipitation with an anionic polymer. Bioseparation 1993; 3:333-341.
43. Guoqiang D, Lali A, Kaul R et al. Affinity thermoprecipitation of lactate dehydrogenase and pyruvate kinase from porcine muscle using Eudragit bound cibacron blue. J Biotechnol 1994; 37:23-31.
44. Larsson EL, Mattiasson B. Isolation of concanavalin A by affinity precipitation. Biotechnol Tech 1994; 8(1):51-56.
45. Larsson EL, Garg N. Kaul R et al. Evaluation of alginate as a ligand carrier in affinity precipitation. Biotechnol App Biochem 1992; 16:48-56.
46. Dissing U, Mattiasson B. Polyelectrolyte complexes as vehicles for affinity precipitation of proteins. J Biotechnol 1996; 52:1-10.
47. Eggert M, Freitag R. Poly-N,N-diethylacrylamide prepared by group transfer polymerization: synthesis, characterization, and solution properties. J Polym Sci 1994; 803-813.
48. Patrickios CS, Gadam SD, Cramer SM. Block methacrylic polyampholyates as protein displacers in ion-exchange chromatography. Biotechnol Prog 1995; 11:33-38.
49. Galaev Y, Kumar A, Agarwal R et al. Imidazole—A new ligand for metal affinity precipitation. App Biochem Biotechnol 1997; 68:121-133.
50. Chen JP, Yang HJ, Hoffman A. Polymer—Protein conjugates. Biomaterials 1990; 11:625-630.
51. Baltes T, Garret-Flaudy F, Freitag R. Investigation of the LCST of polyacrylamides as a function of molecular prarameters and the solvent composition. J Polym Sci 1999; 37:2977-2989.
52. Federici MM, Chock PB, Stadtman ER. Interaction of cibacron blue F3GA with glutamine synthetase: Use of the dye as a conformational probe. 1. Studies using unfractionated dye samples. Biochemistry 1995; 24;647-660.
53. Chen JP, Hoffman AS. Polymer—Protein conjugates. Biomaterials 1990: 11:631-634.
54. Morris, JE, Hoffman AS, Fisher RR. Affinity precipitation of proteins by polyligands. Biotechnol Bioeng 1993; 41:991-997.
55. Kamihira M, Kaul R, Mattiasson B. Purification of recombinant protein A by aqueous two-phase extraction integrated with affintiy precipitation. Biotechnol Bioeng 1992; 40:1381-1387.
56. Guoqiang D, Kaul R, Mattiasson B. Integration of aqueous two-phase extraction and affinity precipitation for the purification of lactate dehydrogenase. J Chromatogr 1994; 668:145-152.
57. Chen JP, Jang FL. Purification of trypsin by affinity precipitation and combining with aqueous two-phase extraction. Biotechnol Tech 1995; 9(7):461-466.
58. Diamond AD, Hsu JT. Aqueous two-phase systems for biomolecule separation. Biotechnology 1992; 47:90-135.
59. Freitag R. Reversibly water-soluble affinity macroligands for bioseparation. Curr Trends Polym Sci 1998; 3:63-79.

CHAPTER 4

Synthetic Displacers for Preparative Biochromatography

Ruth Freitag and Christine Wandrey

Introduction

Biotechnology has become an important industry with a major impact on the economic development in the highly industrialized countries. Bioproduction has the reputation of being "clean" and environmentally friendly, an advantage which is in part paid for by complex production environments. As a consequence, the isolation (downstream process) of the high-value bioproducts, for example recombinant proteins or DNA vectors intended for gene therapy, tends to be difficult for two reasons. First of all, the product is easily damaged and secondly, many of the contaminants/impurities have a very similar chemical nature as the product. For example, they may also be proteins/DNA molecules or even minor variants of the actual product. Concomitantly, biotech products in general and especially those intended for the use in humans have to comply to very strict regulations concerning final purity and consistency.

At present, the bioindustry relies heavily on chromatography to achieve its separation goals. With the exception of gel filtration (also called size exclusion chromatography), chromatographic separations are based on the distribution by adsorption/desorption or partitioning-processes of the feed components between two phases, one which is placed permanently inside a column ("stationary"/adsorptive phase), and a second, which percolates through the column at a given rate ("mobile" phase). Many types of interaction can exist between the molecules of interest in the stationary the mobile phases. The most important amongst them are electrostatic, hydrophobic, and stereospecific (bioaffinity) interactions. Modern chromatography can exploit even very slight differences in protein structure and composition to enforce a separation, and proteins that differ only in a single amino acid or glycosilation site have been successfully separated.

The separation power of chromatography is not restricted to large biologicals; many other substances may be separated chromatographically. In fact, early on chromatography was a separation principle which was extensively used for small molecule separation in the chemical and pharmaceutical industry. There, however, other unit operations such as rectification and extraction soon replaced chromatography for obvious reasons. By comparison, chromatography is a slow and costly technique, which is difficult to scale up. However, in biotechnology a similar change in basic methodology is unlikely to occur and chromatography with it's "physiological" separation conditions (such as aqueous buffers, low temperatures) will most likely continue to be the method of choice for preparative bioseparation for some time to come.

Synthetic Polymers for Biotechnology and Medicine, edited by Ruth Freitag.

In the meantime, chromatographic separations contribute heavily to the production costs of, for example, recombinant proteins, and the question of how to improve the efficiency of these operations needs to be answered. In principle, many aspects of a given preparative chromatographic separation can be improved. One reason why improvements are only slowly entering practice is the severe time constraints for process development. Many downstream processes contain chromatographic steps which are known to be less than optimal in terms of throughput and efficiency, but accomplish the goal. In the past, this problem could often be ignored since cost considerations were less important than entering clinical trials. However, as competition increases, this is expected to change. A more important obstacle to the use of new approaches to improve product isolation—besides an understandable reluctance to use "non-standard" technologies—lies in the fact that their implementation requires some material which is not (yet) commercially available or still too expensive. One example in this context is displacement chromatography, a chromatographic mode which has many advantages over conventional preparative elution chromatography, but which is handicapped by the lack of suitable displacers and stationary phases.

The Principle of Displacement Chromatography

Chromatography began as a predominantly (semi-)preparative technique. From the forties and fifties onward, however, chromatography became more and more synonymous with (high performance) analytical chromatography, HPLC. As a result an impressive number of high-resolution separation techniques were developed in the elution (isocratic or gradient) mode of chromatography. Since analytical elution chromatography does scale up, to some extent many preparative separations were also developed in the elution mode and other modes of chromatography were largely ignored.

The basic difference between elution and displacement chromatography is demonstrated in Figures 4.1 and 4.2. In elution chromatography (Fig. 4.1) the substances are separated into individual "peaks" which are separated by stretches of "empty" mobile phase. In isocratic elution the mobile phase composition does not change during the run. In gradient elution, on the other hand, the mobile phase composition does change during the separation. In the beginning a buffer composition is used, which favors the binding of all but the very low affinity components. During the run the mobile phase becomes more attractive, until towards the end, all but the substances with extremely high stationary phase affinity are eluted. Gradient elution is often necessary in biochromatography since many biopolymers show an "all-or-nothing" type of binding. For a given set of stationary and mobile phases they either bind strongly to the stationary phase or not at all. As a result, only a change of the mobile phase composition throughout the experiments can force all substances to clear the column. Under analytical conditions this presents no challenge since small columns are used. Gradients are much more difficult to realize in the case of large preparative columns, and step rather than linear gradients are often preferred in that case. Such an approach renders the separation of a multicomponent mixture more complex since theoretically for each component an individual step has to be created.

In displacement chromatography, the substances are resolved in consecutive zones of the pure substances, the so-called displacement train (Fig. 4.2). This takes place under the influence of a "displacer". A typically displacement separation is shown in Figure 4.3. In the first stage the substance mixture (feed) is fed into the column. As will be explained below, strongly nonlinear, i.e., competitive, conditions are wanted in displacement chromatography; therefore a large part of the column's capacity can be exploited during this phase. In stage II, the displacer solution is pumped through the column and the separation takes place. As a rule, the displacer is a substance which binds more strongly to the stationary phase than the feed components. Hence the displacer will compete successfully with any feed component for the binding places.

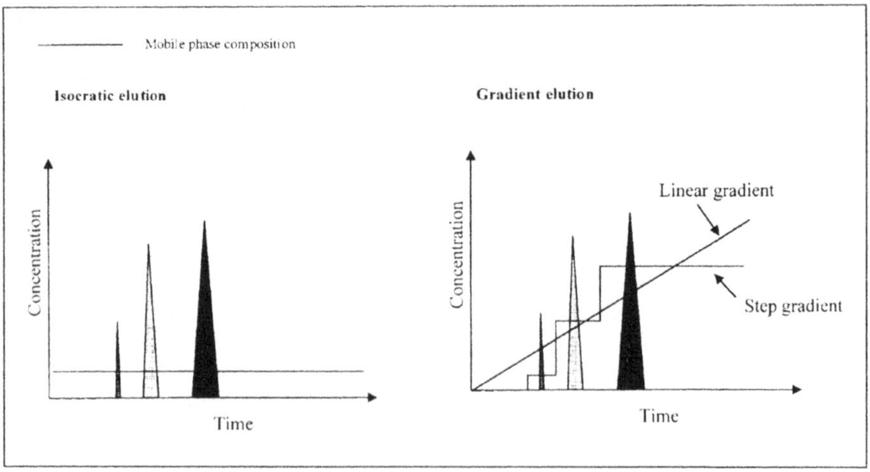

Figure 4.1. Schematic presentation of a separation by (gradient) elution chromatography.

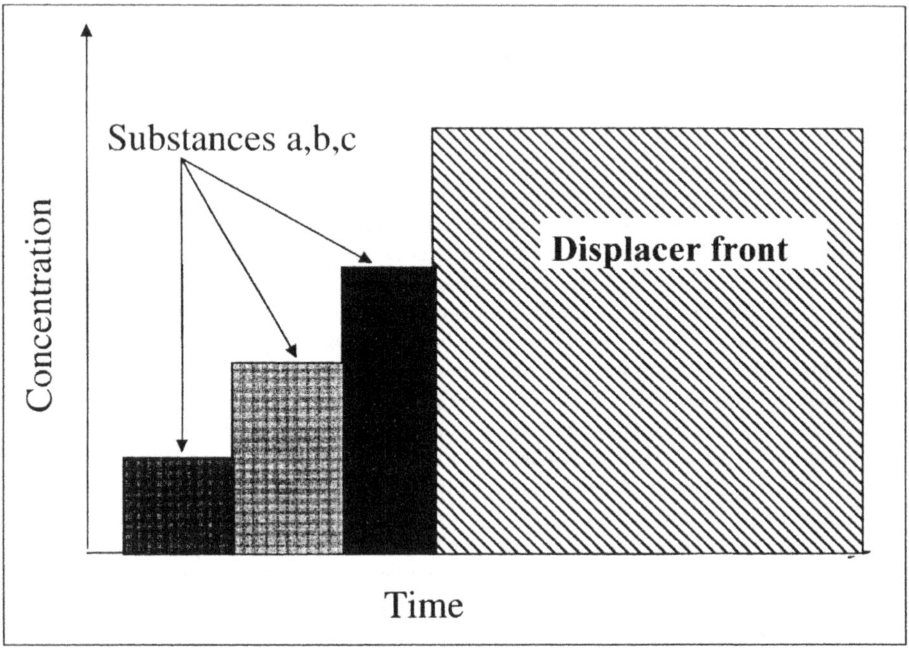

Figure 4.2. Schematic presentation of a separation by displacement chromatography.

A simple "Gedankenexperiment" will elucidate this concept. Imagine a two component system, one feed substance x and the displacer in a situation where the zone development has already taken place. Should a displacer molecule for some reason end up amongst the substance molecules, it will preferably bind to the stationary phase due to its superior affinity and thus be slowed down with regard to the bulk zone of x. Should a molecule of x lag behind the bulk zone of x, it is less likely to become adsorbed than the displacer molecules and therefore will be sped up in relation to the displacer front. If there is more than one component in the mixture, each

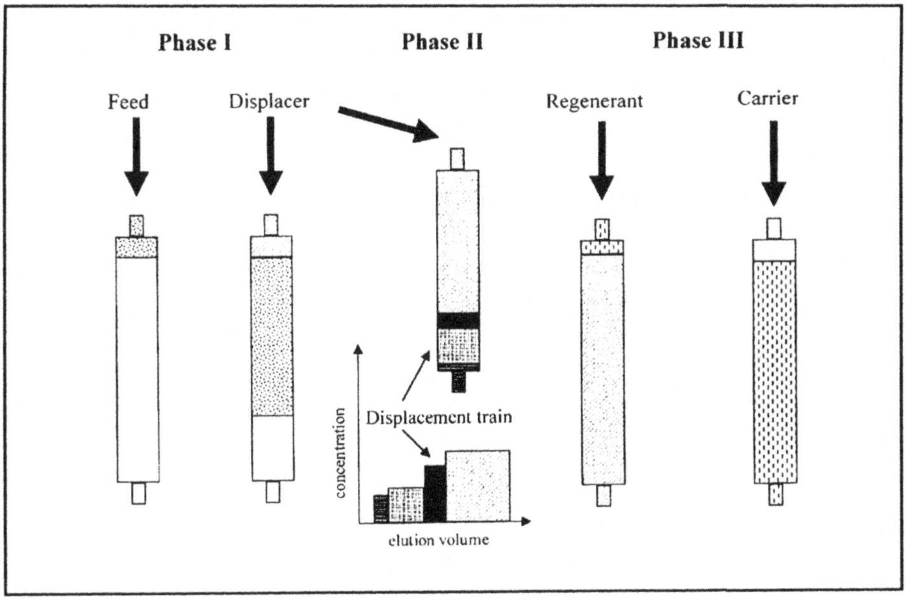

Figure 4.3. The different stages of a displacement separation: stage I, introduction of the feed, stage II, introduction of the displacer and development of the displacement train, stage III, column regeneration and reequilibration.

molecular species will compete for the binding sites. The competition increases during the run, since the number of available binding places is continuously reduced as the displacer front advances. The result is the formation of the consecutive zones of the pure substances as shown in Figure 4.2. It is important that in displacement chromatography the displacer always stays behind the substance zones, while in elution chromatography the gradient is superimposed over them. Moreover, a single displacer step is sufficient to resolve a multicomponent mixture.

Some simple mathematical considerations with regard to the development of the displacement train are given in Insert 4.1. Once the displacer concentration in the carrier ($c_{Displacer}$) has been adjusted, the corresponding stationary phase concentration of the displacer ($q_{Displacer}$) follows from the displacer isotherm. The q/c ratio is a constant for all substances within the system, equation 3, so if the isotherm of a substance a is known, the concentrations c_a of this substance in the displacement train can be easily determined. An "operating line" can be constructed by drawing a straight line from the origin of the plot to the point on the displacer isotherm corresponding to c_d (Fig. 4.4) top. The concentrations of the individual substances in the displacement train can be taken from the intersection point of the operating line with the substance's isotherm. Substances whose isotherms do not intersect with the operating line elute ahead of the displacement train (Fig. 4.4 bottom).

This is an oversimplified presentation of the displacement separation which does not take into account important aspects, such as mass transfer restrictions, kinetic and extra column effects, nonLangmurian isotherms (a common occurrence in biochromatography) or changes in the local microenvironment of the substances. Protein adsorption is strongly influenced by such local changes in the mobile phase, as will be discussed below. For theoretical considerations most of the above-mentioned effects can be taken into account (for an introduction to the theory of displacement chromatography, see e.g.[1,2]). However, the simplified picture already allows some insight into the advantages and possibilities of displacement chromatography.

Insert 4.1.

The speed of a substance front moving through a chromatographic column, u_l, depends on the energy of adsorption (as indicated by the isotherm $q_l = f(c_l)$), the flow rate of the mobile phase, u_o, and the phase ratio, ϕ [65].

$$\mu_i = \frac{\mu_o}{\left(1+\phi \, \partial q_i / \partial c_i\right)} \qquad \text{Eqn. 1}$$

In the displacement train all substance zones have to move at the speed of the displacer front and therefore equation 1 has to apply all of them, i.e.,

$$\frac{u_o}{(1+\phi \, \partial q_D / \partial c_D)} = \frac{u_o}{(1+\phi \, \partial q_a / \partial c_a)} = \frac{u_o}{(1+\phi \, \partial q_b / \partial c_b)} = \ldots \qquad \text{Eqn. 2}$$

With subscript D indicating the displacer and subscripts a and b indicating individual substances. Since mobile phase flow rate, u_o, and phase ratio, ϕ, are constants, it follows that:

$$\frac{dq_a}{dc_a} = \frac{dq_b}{dc_b} = \frac{dq_c}{dc_c} = \frac{dq_{Displacer}}{dc_{Displacer}} = \frac{d_{Displacer}}{c_{Displacer}} \qquad \text{Eqn. 3}$$

Since the displacer concentration determines the concentrations in the displacement train, the feed concentration is of little importance and high concentration factors can be achieved. In some cases the possibility to control the concentration of a given substance can also be of advantage, for example, if aggregation or denaturing occur at high concentrations. Since only a single displacer step is needed to separate a multicomponent mixture, displacement chromatography lends itself more easily to preparative applications than step elution chromatography. Some recent comparisons have indeed shown that significant savings in terms of costs, time and buffer consumption are possible if the displacement rather than the elution approach is used under otherwise identical conditions.[3,4] The advantages of displacement chromatography are especially prominent when expensive feeds are to be processed (oligonucleotides, synthetic peptides) or for the final polishing step of a given product. In this case all remaining impurities will be focused into small but highly concentrated zones adjacent to the zone of the product. Their removal is therefore possible with a minimum of product loss. Another application, which evolved only recently, is the utilization of displacement chromatography for "sample preparation" in analytical chemistry, i.e., the direct coupling of the displacement column to an analytical system such as a mass spectrometer.[5] Again, the major advantage is the focusing/concentration of minor and even trace components of the sample. The detection limit of the hyphenated system is thereby improved considerably in comparison to that of the detector alone.

In spite of these acknowledged advantages and in spite of some encouraging cases in the recent past, displacement is rarely applied in today's bioindustry. A major problem is the displacer, since this substance is the basic requirement of any displacement separation. While finding a

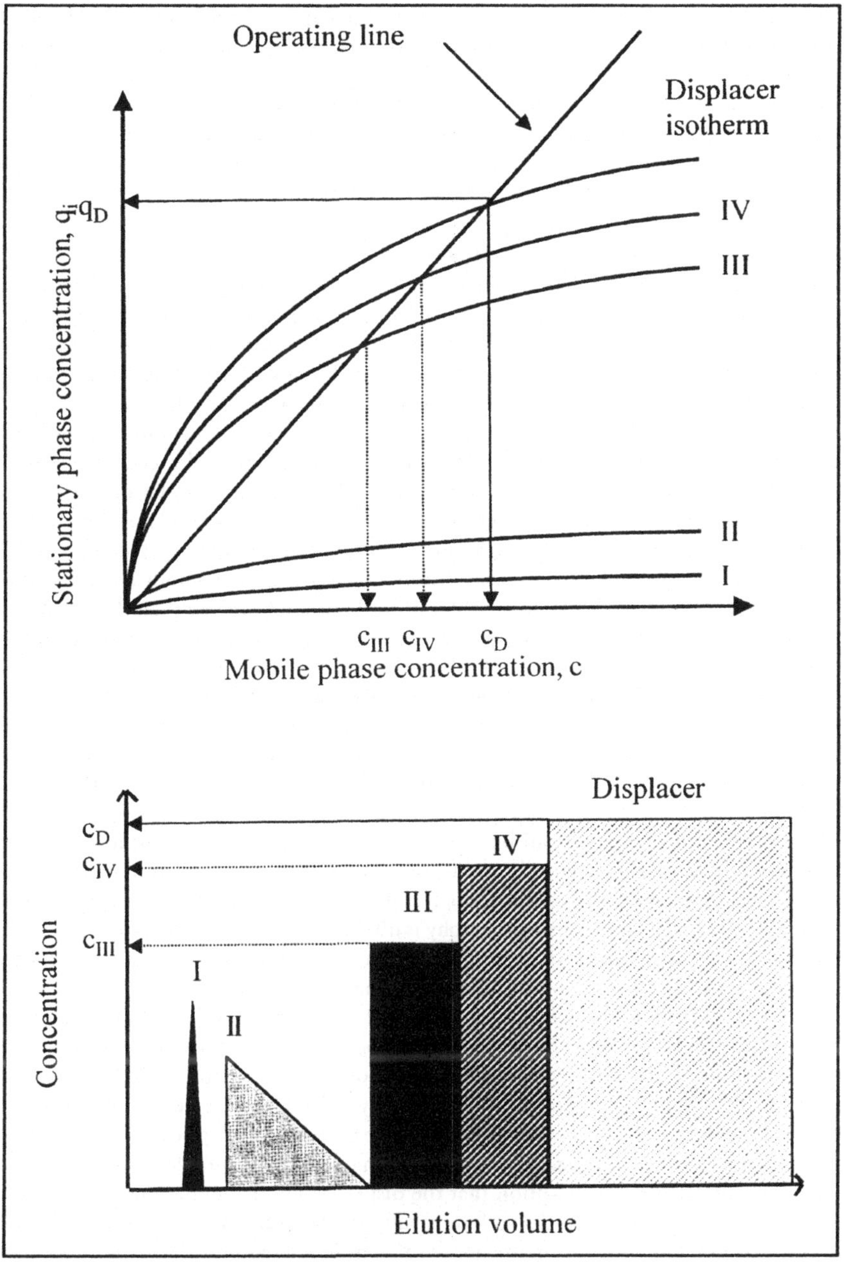

Figure 4.4. Simulation of the displacement train under ideal conditions from a set of individual substance isotherms and the displacer isotherm. The operating line is found by connecting the origin of the plot with the point on the displacer isotherm, defined by the displacer concentration in solution. The intersection points of the operating line with the substance isotherms determined the substance concentration in the displacement train. Substances whose isotherms do not intersect with the operating line elute ahead of the displacement train. Top: Multiisotherm-plot with operating line. Bottom: Corresponding chromatogram (displacement train).

suitable substance is still comparatively easy for the separation of small molecules, it becomes the major challenge in the separation of (bio)macromolecules. The physicochemical properties of the displacer determine the quality and feasibility of a separation. However, our understanding of how certain (measurable) physicochemical displacer properties are linked to the observed chromatographic behavior is poor. Manipulation of certain chemistry-linked undesirable qualities such as displacer toxicity, difficult column regeneration or low mobile phase solubility becomes purely empirical under these circumstances.

Displacers for Biochromatography

For application in biochromatography a displacer has to fulfill some basic requirements. It should, for example, be nontoxic, biocompatible, stable and cheap. The latter is especially important, since large amounts of displacer are needed and cost considerations often determine whether or not a particular procedure is used in industry. In this context the possibility of displacer recycling should be investigated and validated, since this can become a considerable asset in terms of lowering the production costs. After the separation it must be possible to remove the displacer from the final product to the required degree and to assess the residual displacer concentration with a sensitive and selective assay. For pharmaceutical applications it might even be necessary to sterilize the material. The ideal displacer should dissolve well in the mobile phase since high displacer concentrations result in fast separations and high product concentrations. Stationary phase affinity is already an ambiguous issue. It has been said before that the displacer's stationary phase affinity should be higher than that of any of the substances to be displaced. However, column regeneration must also be considered and some authors suggest that fine-tuning the affinity rather than going for the highest possible one actually yields better results in terms of resolution, yield and throughout.[6]

Displacement biochromatography has in the past been predominantly a method for the separation of small molecules.[7-8] Amino acids, peptides and even small proteins such as insulin were separated in the reversed phase mode, using hydro-organic or aqueous mobile phases and hydrophobic, but soluble, small molecules as displacer, such as 2-(2-butoxyethoxy)ethanol (BEE), decyltrimethylammonium bromide, cetyltrimethylammonium bromide (cetramide), benzyldimethyldodecylammonium bromide, benzyldimethylhexyldecylammonium chloride, benzyltributylammonium chloride, dodecyloctyldimethylammonium chloride and palmitic acid. Unfortunately, reversed phase chromatography is usually not suited to preparative protein chromatography since many proteins are damaged by the hydro-organic solvents and used and hydrophobic stationary phases in this type of chromatography.

For various reasons, ion exchange displacement chromatography (IEDX) using polyelectrolytes as displacer has become the predominant mode for protein displacement chromatography and other modes such as hydrophobic interaction chromatography (HIC), affinity chromatography (AC) and hydroxyapatite chromatography (HAC) have been used only occasionally.[8] In the classic applications two types of molecules have been used as displacers of large biologicals in IEDX; these were proteins and high molecular mass synthetic polyelectrolytes. The reason for this restriction was the assumption that the displacer of a (large) protein molecule would need to be fairly large itself in order to have a sufficiently high affinity for successful binding place competition. Early reports on the use of small molecules as protein displacers were disregarded due to the heterogeneity of the displacer substance used.[9] It was assumed that high molecular mass impurities were the actual displacers. However, it was suggested as early as 1990 that low molecular mass molecules may act as protein displacers since proteins are known to use only a fraction of their surface to interact with an ion exchanger as a consequence of their surface charge asymmetries. We know today that such small displacers have advantages in terms of column regeneration and fine-tuning of the separations, and recently methods for the synthesis of high-affinity low molecular mass displacers have been proposed which combine the advantages of the conventional small displacers with those of the classical high molecular mass molecules.[10-11]

The early applications of displacement chromatography were all in the area of protein purification. However, lately the isolation of DNA is gaining importance since larger amounts of highly purified plasmid DNA are, for example, needed in gene therapy. As opposed to proteins, DNA molecules are themselves large and linear polyelectrolytes. As the first applications of displacement chromatography to the purification of plasmid DNA are being published, the question of what type of molecule constitutes a suitable DNA displacer may give a new impetus to the development of the theory of displacer design.[12] Up to now, however, there is no systematic way to design a displacer's physicochemical properties (size, structure, type and density of specifically and nonspecifically interactive groups) with the aim of achieving sufficient stationary phase affinity for a given separation. In order to do this, it will be necessary to advance our understanding of the solubility and adsorption of large and small molecules and not only of polyelectrolytes. Displacement chromatography will not be the only area to profit from this. In a first attempt to classify displacers according to physical properties, which is known to influence the chromatographic behavior, we propose nine different classes of displacer as indicated in Table 4.1.

Classical Protein Displacers

The lack of a suitable displacer can easily prevent an entire section of the stationary phase/interaction spectrum of chromatography from being used in the displacement mode. For example, hydrophobic interaction chromatography (HIC), the preparative alternative to reversed phase elution chromatography in the protein area, has not gained the same importance in displacement chromatography.[13,14] To date, protein displacement chromatography is almost exclusively performed in the ion exchange mode (IEDX). Typical displacers in these applications are (modified) natural compounds such as chondroitin sulfate, dextran sulfate, carboxymethyl starch and alginate or purely synthetic ones such as Eudragit, Nicolyte 7105 and poly(ethylene imine) (PEI). These substances hence fall in the P (n,m,s) category of displacer (Table 4.1).

Synthetic polymers are cheap, available in large variety and can often be sterilized (e.g., by heat or filtration). However, they are generally only available in heterogeneous mixtures varying considerably in molecular mass and structure (e.g., branched versus linear chains). A measure for the molecular mass distribution is the index of polydispersity, which would be 1 in the case of no variation within the molecule population (e.g., for a protein), below 1.5 for a typical "homogeneous" synthetic polymer and easily above 10 for most of the commercially available synthetic polymers. Displacer heterogeneities and impurities pose a grave problem since the stationary phase affinity of the different molecules varies considerably and often less well retained displacer molecules contaminate the product zones.[15] In addition, the highly charged high molecular mass fractions of the displacer preparation tend to have a very high stationary phase affinity and are therefore difficult or impossible to remove from the column. Solubility tends to be restricted for large polyelectrolytes and many of the synthetic polymer preparations contain toxic impurities (e.g., residual monomer molecules). The high viscosities of concentrated polymer solutions tend to pose additional problems in the utilization of P-type protein displacers.

Proteins themselves can also be used as protein displacers. They have the disadvantage of being expensive and fragile and are generally not suited for large-scale pharmaceutical applications. However, they also have some advantages over the above-mentioned synthetic polymers since they represent a homogeneous molecule population and are susceptible to detection by UV/VIS or specific bioassays. Established rules concerning how to deal with possible residual protein contamination of a given product exist. In addition, any interaction mode of protein

Table 4.1. Classified protein displacers

	Low Molecular Mass Displacer (L)	Oligomeric Displacer (O)	Polymeric Displacer (P)
Natural (n)	Ln	On	Pn
Modified natural (m)	Lm	Om	Pm
Synthetic (s)	Ls	Os	Ps

chromatography can be used with proteins as protein displacers, since it will almost always be possible to find a protein which binds more strongly to the column in question than the target molecule. One of the earliest displacement separations in the HIC mode, for example, used a protein (bovine serum albumin) as displacer.[13]

Low Molecular Mass and Oligomeric Protein Displacers

Until the early 1990s, the idea of low molecular mass displacers for ion exchange displacement chromatography seemed absurd. However, isolated cases of the utilization as well as speculations concerning the putative advantages and disadvantages of such substances continued to appear in the pertinent literature. Jen and Pinto were amongst the first to report on the use of a Ls-type substance, namely poly(vinylsulfonic acid) (Mw 2000 g/mol) as protein displacer.[9] Concluding from their investigation of dextran sulfates (Om- / Pm-type) as protein displacers the same authors clearly speak in favor of using the Om- (Mw <5000 g/mol) rather than the Pm-variant (Mw > 500'000 g/mol) for that purpose.[16] No separation of the test mixture, α-lactoglobulins A and B, could be achieved with the Pm-type displacer. It was speculated that such extremely large displacer molecules are excluded from certain pores and are thereby prevented from acting as displacer in spite of their potentially high affinity (the number of charges per molecules is 6100 for the Pm versus <100 for the Om analogs of similar chemistry). The Om-type molecules had the additional advantage of being nontoxic (dextran sulfates with a molecular mass of <10,000 g/mol having been administered to humans orally and intravenously without adverse results) and easily detected (strong UV absorbance at 260 nm). They can be sterilized, show no nonspecific interaction with proteins and are strongly ionic and cheap. Using this system, a first investigation of the influence of the molecular mass on the displacer properties was carried out. It was shown that the fairly small molecules can act as displacer and that column regeneration is easier in their case. Especially the latter advantage was among the first to be noticed for small versus large displacer molecules.

In one of the few cited cases of affinity displacement chromatography, Ls-type poly(ethylene imine) molecules, such as PEI-2000, PEI-700, were better displacers than the Os-type PEI-35'000.[17] Spermine, a low molecular mass analog (degree of polymerization of 4, Mw 202.34 g/mol) of PEI, on the other hand, did not act as displacer. Small calcium chelating molecules such as IDA (imminodiacetic acid, Mw 133.4 g/mol) and EGTA (ethyleneglycolbis (β-aminoethylether)-N,N,N'N'-tetraacetic acid, Mw 380.4 g/mol) were used as protein displacers in hydroxyapatite and anion exchange chromatography.[4,18] Beginning in the early 1990s, the group of Cramer has also reported repeatedly on the use of L- and O-type molecules as protein displacers, including, for example, protected amino acids, antibiotics, heparin, and various sulfonic acids such as: 1,2 benzene sulfonic acid (BDS), 1,5 naphthalene sulfonic acid (NDS), and p-toluene sulfonic acid (PTS).[2]

As the understanding of how displacers enforce the separation advance, it became evident that small molecules can—under the right circumstances—indeed constitute powerful protein displacers. L- and O-type displacers are more sensitive to the actual chromatographic conditions and in the case of IEDX most specifically to the salt content of the mobile phase than P-type displacer molecules. However, they have the advantage of a more uniform molecular mass and hence more homogenous stationary phase affinity. They can easily be removed from the protein product by filtration or dialysis. O- and especially L-type displacers are often cheaper than P-type ones. They have the additional advantage of superior mass transfer properties and even at high concentration they do not increase the viscosity of the mobile phase to the same extent as a polymer would. Increasingly such small molecules are used and important deduction are drawn from the results. A number of these small protein displacer molecules and their application are listed in Table 4.2.

Design of Protein Displacers

Torres and Peterson were the first to try to synthesize a dedicated displacer. Starting in 1978, they chemically modified high-molecular mass carboxymethyldextranes (CM-D) to obtain a series of molecules with graded affinity to anion exchange materials.[19] The mixture behaved as a single displacer as long as all displacer components had a higher binding affinity than the target molecules. If this was not the case, the low affinity fractions of the "displacer" would penetrate into the displacement train and position themselves between protein zones of lower and higher affinity respectively. It was possible to remove the spacer/displacer molecules from the protein fractions by hydrophobic interaction chromatography if necessary since normally only proteins interacted with such columns, but not the CM-D.[20] Since the CM-D are nontoxic and non-UV active this spacer-potential may be considered as an advantage. The low affinity CM-D variants separate the individual protein zones and thereby facilitate their detection (spacer displacement chromatography).[21] It must be acknowledged, however, that the high preparative potential of displacement chromatography is realized fully only if target molecule-target molecule displacement is exploited to create the displacement train.

In the early eighties, CM-D fractions with fairly narrow ranges of affinity became available and these substances were commercialized.[22,23] A system was developed to give an "index of affinity" to each of the CM-D fractions.[22] The weak anion exchanger DEAE-Sephadex A-50 shrinks when saturated with CM-D. The volumetric change is a function of the A220/Δn index of the CM-D solution used for the saturation and a linear relationship exists between the reciprocal of the pellet volume (RPV) and this index. The A220/Δn ratio can be calculated from the absorbency at 220 nm of a 3% CM-D solution in water divided by the difference in the refractive index (n) of this solution and water. The shrinking of the DEAE Sephadex beads upon saturation with CM-Ds is most likely due to a reduction of the electrostatic repulsion of the DEAE surface groups. Since at saturation approximately the same number of surface charges are neutralized by the CM-D, the RPV characterizing the CM-D with respect to its relative charge density, i.e., its relative content of carboxyl groups. This allows assigning a "relative affinity" to CM-D having a specific molecular mass. The respective molecules can subsequently be used as displacer (or spacer) in a given separation, the affinity of each CM-D molecule being a direct function of the number of negative charges it can present to the adsorbent. CM-D are relatively inexpensive materials. They are produced under controlled conditions and cover a wide affinity range. Heterogeneous mixtures of CM-D were used to separate such complex samples as guinea pig serum, mouse liver cytosol, and alkaline phosphatase from *E. coli* periplasm.[24]

Following these early efforts, the year 1995 can be considered as the starting point of rational displacer design, since at least three different types of synthetic displacer of the Ls- and Os-type were introduced in that year. Breier and Freitag introduced a series of poly(ethylene

Table 4.2. Small protein displacer molecules

Displacer	Target molecule / Conditions	Comments
Dextran sulfates (Mw < 5000 g/mol)	Standard protein mixture/anion exchange DC	The smaller molecules are the better displacers [16]
Pentosan polysulphate (Mw 3000 g/mol)	Whey proteins/anion exchange DC	Comparison to dextran sulfate (Mw: 50'000 g/mol) [61]
Pentaerythritol based dendritic polymers (Mw: 480—5100 g/mol)	Standard protein mixture/cation exchange system	SMA model, steric factor and charge density more important than size [62]
PEG based linear and dendritic polymer carrying end chelating groups (Mw: 1000-50'000g/mol)	Standard protein mixture/hydroxyapatite chromatography	Smallest molecule best results [25]
EGTA (Mw: 380.4 g/mol) and IDA (Mw: 133.4 g/mol)	Standard protein mixture/ hydroxyapatite chromatography	

continued on the next page

Table 4.2. Cont.

Poly(vinylsulfonic acid) (Mw 2000)	Ovalbumin/conalbumin-separation, fractionation of ovalbumin/ anion exchange DC	L-type displacers superior to P-type ones [9]
Protected amino acids (Arginine, Lysine esters)	Standard protein mixture/cation exchange DC	[10]
Aminoglycosidic antibiotics (Neomycin B, Streptomycin A)	Strongly adsorbing proteins (lysozyme)	[11]
Variety of L-type displacers	Standard protein mixture/anion exchange chromatography	L-type displacers depend more on the conditions (concentration buffer composition) than high Mw analogs [63]
Heparin	Standard protein mixture/anion exchange chromatography	Nontoxic, homogeneous Ln-type displacer [64]

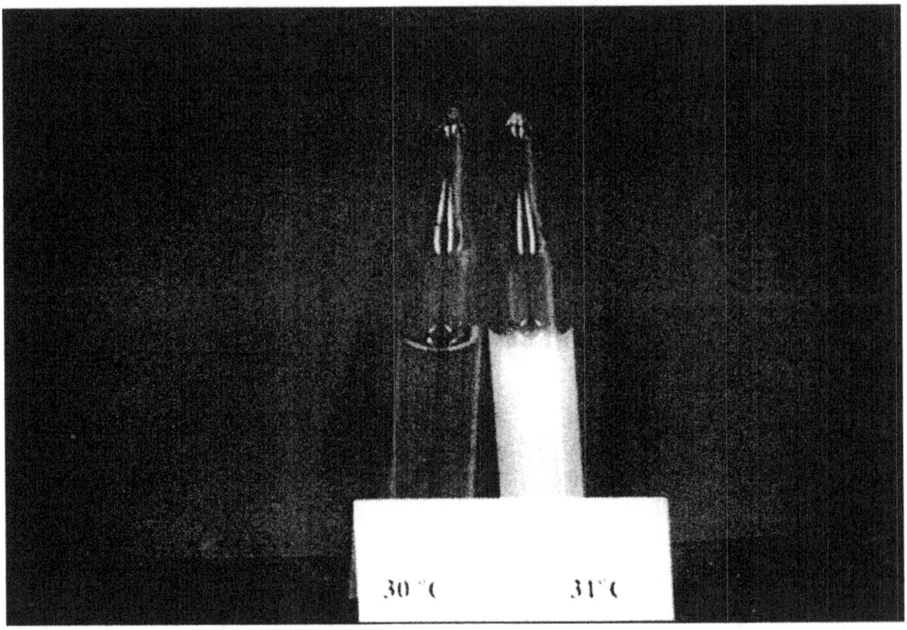

Figure 4.5. Thermosensitive displacer having a lower critical solution temperature between 30°C and 31°C.

glycol) based linear and dendritic polymers (Mw between 1000 and 50'000 g/mol) that were modified to carry chelating groups at the ends.[25] These molecules were used as protein displacers in anion exchange and hydroxyapatite chromatography. Vogt and Freitag first introduced their building block method for target-orientated displacer synthesis by copolymerization.[26] This approach allows the production of a displacer for a given stationary phase and/or separation problem. The stationary phase determines the nature of a so-called specific monomer, such as ionic groups for ion exchange materials. If useful, a second monomer is used to add features to the displacer, which, for example, facilitate recovery or detection. In a first application this method was used to produce a thermo-precipitable displacer, which was soluble in water at temperatures below a certain critical temperature, but rapidly and quantitatively precipitated if this temperature was surpassed by even the fraction of a degree (Fig. 4.5). The precipitation/resolution was very fast and the displacers could be run through this cycle several hundred times. The tendency for nonspecific coprecipitation, for example, of proteins could be kept low.

A somewhat similar approach was been taken by Patrickios et al in collaboration with the group of Cramer.[26] In this approach, the technique of Group Transfer Polymerization was used to produce di- and tri-block polymethacrylates bearing a sequence of positively charged groups at one end, a sequence of negatively charged ones at the other, and a neutral, hydrophobic middle block. Due to their polyampholytic nature, such polymers show isoelectric points much like proteins. The authors claim, that their polymers are suitable protein displacers for anion but also cation exchange displacement chromatography depending on the relative ratio of positively and negatively charged groups in the molecule and the pH of the carrier, while the ampholytic nature of the molecules facilitates recovery, e.g., by precipitation at the displacer's isolelectric point.

Quintero et al proposed in 1995 a method for the dedicated design of a series of homologous generic displacers for chiral stationary phases.[28] The corresponding stationary phase was α-cyclodextrin activated one (Pirkle-type, Cyclobond II). The displacers (Fig. 4.6) contain an

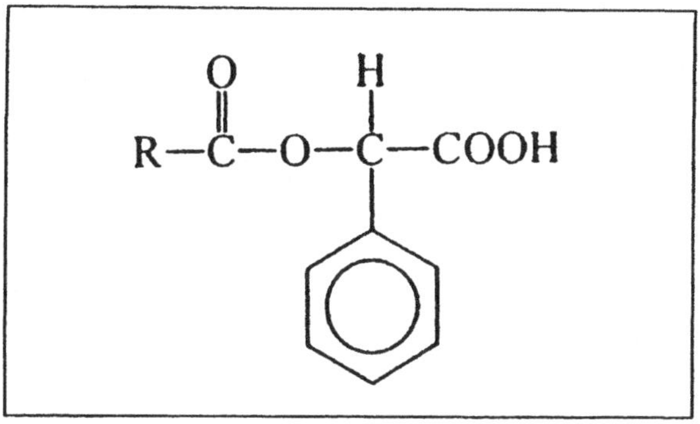

Figure 4.6. Displacer designed for chiral selective stationary phases (Cyclobond II). The tail section R can vary in length and serves to adjust the solubility. The phenyl group fits into the cavity of the α-cyclodextrine, and the carboxyl and carbonyl groups serve to form multiple hydrogen bonds with the secondary hydroxyl groups of α-cyclodextrin [reproduced with permission in modified from reference 28].

anchoring phenyl group, which fits in the cavity of α-cyclodextrine. A second section of the molecule carries carboxyl and carbonyl groups that may form multiple hydrogen bonds with the secondary hydroxyl groups of α-cyclodextrin. Finally, a solubility-adjusting tail section (alkanoate group with variable chain length) is present, which regulates the displacer's solubility. The retention and adsorption properties of these displacers were found to be a function of the size of the alkanoic group. Longer side chains result in higher capacity factors and increased retention.

In 1998 Ruaan et al suggested methacrylic block copolymers similar to the ones proposed by Patrickios as protein displacers in hydrophobic interaction chromatography on octyl-sepharose.[14] The molecular masses of the displacer were relatively small (between 3000 and 4000 g/mol). When the isotherms were recorded, the displacer isotherms were below the protein (trypsin, α-chymotrypsin) isotherms (Fig. 4.7). According to the conventional formalism, this would mean that the triblock polymers could not function as displacer for these proteins. However, in the experiment these substances did work and the authors proposed a modified Langmuir model which introduces a "hindrance factor", δ, which is negligible for the (small) displacer but significant for the proteins. As a result the isotherm system can be adjusted to appear "normal". However, the exact physicochemical basis of this hindrance factor remains hidden.

DNA Displacers

Nucleic acids such as DNA and RNA constitute a second important group of biological macromolecules besides proteins. While proteins have for some time now been isolated preparatively in fairly large scales, DNA purification has only recently entered that realm. Compared to proteins, which are constructed from more than 20 amino acids of varied hydrophobicity and which in addition can contain complex sugars, lipids, or even metal ion, the structure of the DNA molecule is much simpler. From a chromatographic point of view, DNA can be considered as a very large linear polyanion with a very homogeneous charge distribution. Both anion exchange and hydroxyapatite (interaction with the phosphate backbone) chromatography can be used for its purification. Nevertheless, DNA also presents some specific challenges to large-scale preparation. Concentrated DNA solutions are very viscous. Especially in elution chromatography, this causes problems since the peak maximum should stay below the viscosity/concentration limit. As a result only low concentration preparations can be obtained. In this context, displacement chromatography has the theoretical advantage of allowing to keep the concentration in the entire zone just below the viscosity/concentration limit. The overall concentration of the "product fraction" will be higher in this case.

According to a recent publication, the separation of protein and plasmid DNA as well as that of DNA and lipopolysaccharides is possible by displacement chromatography.[12] Linear

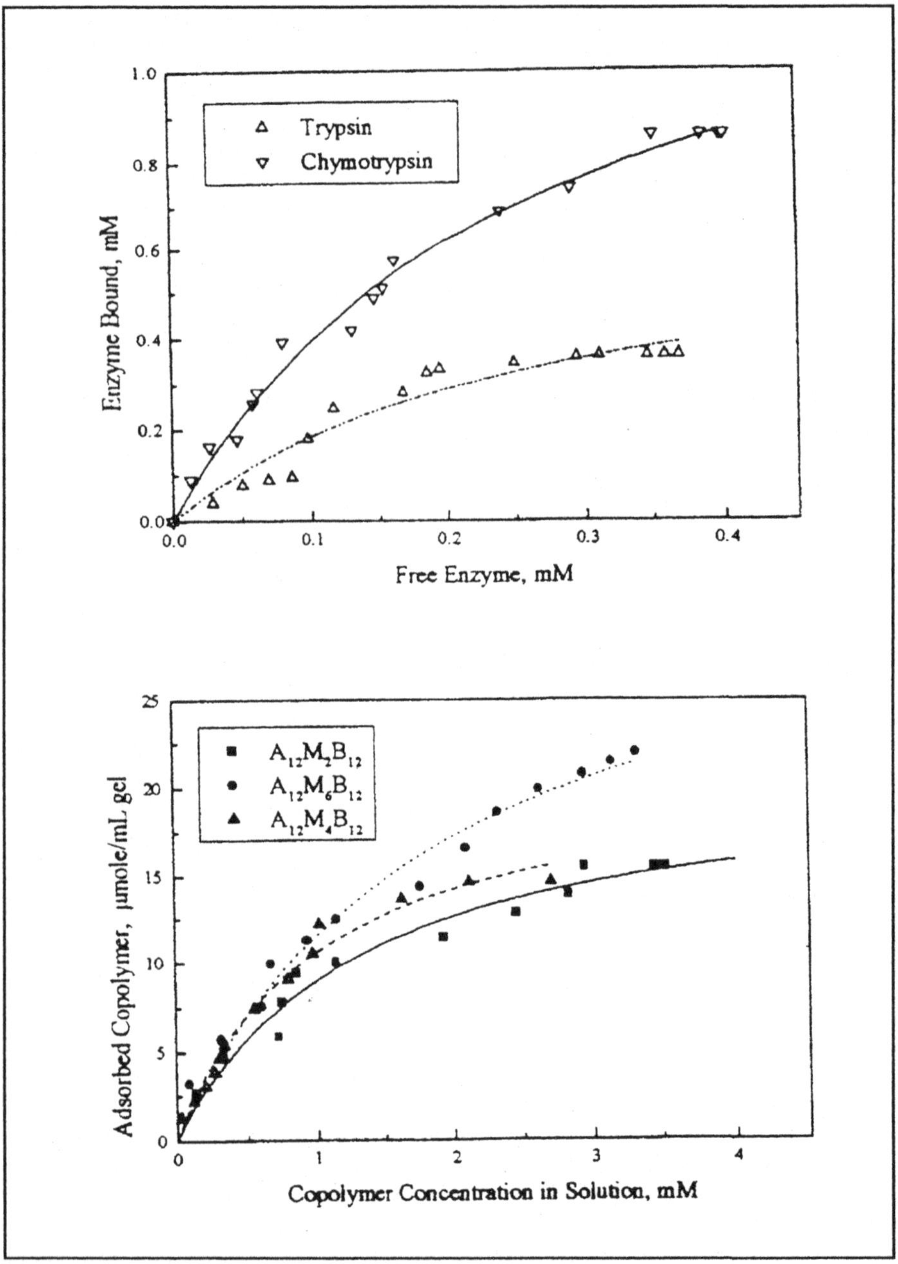

Figure 4.7. Single-component isotherms of (top) proteins and (bottom) synthetic tri-block copolymers measured at room temperature for Octyl-sepharose stationary phase. The affinity constants calculated from an isotherm fit according to the Langmuir model seemed to indicate that the proteins bind more strongly than the copolymers and hence could not be displaced by them [reproduced with permission from reference 14].

polyacrylic acid (Os-type, Mw 5100 g/mol) was successfully used as displacer. A particularity was the fact that conventional stationary phase materials based on porous anion exchanger or hydroxyapatite beads could not be used for the separation of the plasmid DNA from a standard protein (trypsin). The use of a monolithic column (UNO Q1, BioRad), on the other hand, yielded promising results. In this case protein and DNA capacities were comparable. Further research is obviously necessary. However, the investigation of DNA displacement also offers some principal opportunities, for example, to elucidate the effect of size and rigidity, or charge density and chemistry for the competitive adsorption of linear polyelectrolytes.

Polyelectrolytes

Several definitions and classifications for polyelectrolytes have been published. Herein, we will define a simple polyelectrolyte as a homopolymer carrying an ionizable group at each monomer unit (Fig. 4.8). Such homopolymers may be strong polyelectrolytes; in this case their ionic groups are strong bases or acids and their charge is nearly independent of the pH. In contrast, weak polyelectrolytes carry weakly basic or acidic groups and their solution behavior strongly depends on the pH. More complex polyelectrolytes are, for example, copolymers where only a fraction of the monomeric units carries charges, and polyampholytes, where both positive and negative charges are distributed along the polymer chain. Proteins may in this context be classified as very complex polyampholytes.

Solubility of Polyelectrolytes

Generally, the solubility of a polymeric substance in a solvent is determined by the interaction solute/solvent relative to the interaction solute/solute and solvent/solvent. However, the number of molecular and segments conformations in solution, relative to the pure substance, also influences the solubility. In the case of polyelectrolytes these interactions can be of different types. Although Coulombic interactions play the dominant role in aqueous polyelectrolyte solutions, other intermolecular forces including van der Waals, hydrophobic, hydrogen bond and dipol-charge transfer also have to be considered to a varying degree as factors for the solubility. Figure 4.8 shows a schematic of a simple polyelectrolyte chain and summarizes the physical parameters, which can influence the solubility. These are in particular the chain length (molar mass), the charge distance (charge density), and the distance between the polymer backbone and the charged groups (spacer length). However, the chemical nature of the ionic group, the substituents at the ionic group and the counterions also influence strongly the solubility. Other factors can include the structural nature such as the chain architecture (branching, cross-linking) or the charge density distribution along the polymer chain (homogeneous, heterogeneous).

For an evaluation of the behavior of a polyelectrolyte in solution, the charge density along the polymer chain has the most significant effect, since it determines among other things the distribution of counterions around the polyion. For low charge densities, characterized by a charge distance much higher than the so-called Bjerrum length (0.71 nm in water at 293 K), the counterion distribution is within a "Debye-Hückel" regime.[29-32] If the charge distance is less than the Bjerrum length, the electrical field around the polyion becomes so strong that some of the counterions become trapped within very short distances from the polyion.[33-35] The superposition of the aforementioned features ultimately governs the solution behavior of the polyelectrolytes. The solubility of a polyelectrolyte in an aqueous system is thus largely determined by the ionic strength of the system.[36] In general, an increase of the ionic strength decreases the solubility, primarily due to a change of the conformation of the polyelectrolyte via screening effects as well as a thermodynamically controlled salting out effect by deterioration of the solvent quality. Both effects are superimposed and determine the state of solution.

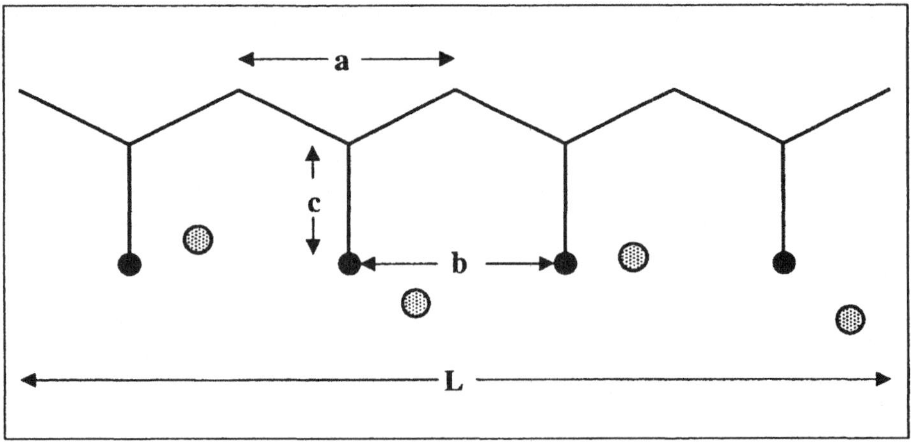

Figure 4.8. Schematic presentation of a simple polyelectrolyte chain. • chain charge, counterion, a: monomer unit length, b: charge distance, c: spacer length, L: contour length.

Furthermore, at a defined ionic strength, the kind of counterions, their valency and their state of hydration can be significant. In particular, for monovalent counterions the hydration is crucial. Divalent or multivalent counterions can act as centers of cross-linking via intra- and/or intermolecular bridges.[37]

Recently it could be experimentally shown that the ratio of the Debye length to the contour length is a crucial parameter in order to identify changes in the polyion/counterion interaction.[38] Therefore, the influence of the molar mass on the solubility has always to be evaluated for a given ionic strength resulting from the polyelectrolyte and salt concentration. This becomes very important for the application of polydispersed polyelectrolytes. In this case an increase of the ionic strength can result in a partial aggregation of the longer chains whereas the shorter chains remain completely in solution. At present the solubility of a polyelectrolyte in a given solvent cannot be predicted from molecular data. However, it can be evaluated experimentally, for example, by measuring the intrinsic viscosity in the solvents of interest. The stronger the interaction with the solvent the higher the intrinsic viscosity will be. For all applications of polyelectrolytes in aqueous, ion-containing media, changes of the solubility are expected to influence the final properties.

Adsorption of Polyelectrolytes

The term adsorption denotes generally the increase in concentration of a solute in an interfacial region. Adsorption is used for different purposes. Since the adsorbed molecules can give the interface a number of interesting properties, adsorption is a method for a broad variety of material/surface modifications. On the other hand, adsorption can also be an intermediate step in separation processes such as chromatography. The adsorbed amount is a key parameter for the description of the adsorption process. The adsorbed amount is normally plotted as a function of the equilibrium bulk substance (polymer) concentration at a given temperature (adsorption isotherm). Figure 4.9 shows a high-affinity polymer adsorption isotherm, which is typical for a monodisperse polymer (solid line) and a second isotherm more typical for a polydisperse polymer (broken line). As it can be seen, a high degree of adsorption occurs even at low bulk concentrations in the case of monodispose polymers. For higher bulk concentrations the attainment of a plateau indicates the saturation of the surface (monolayer).

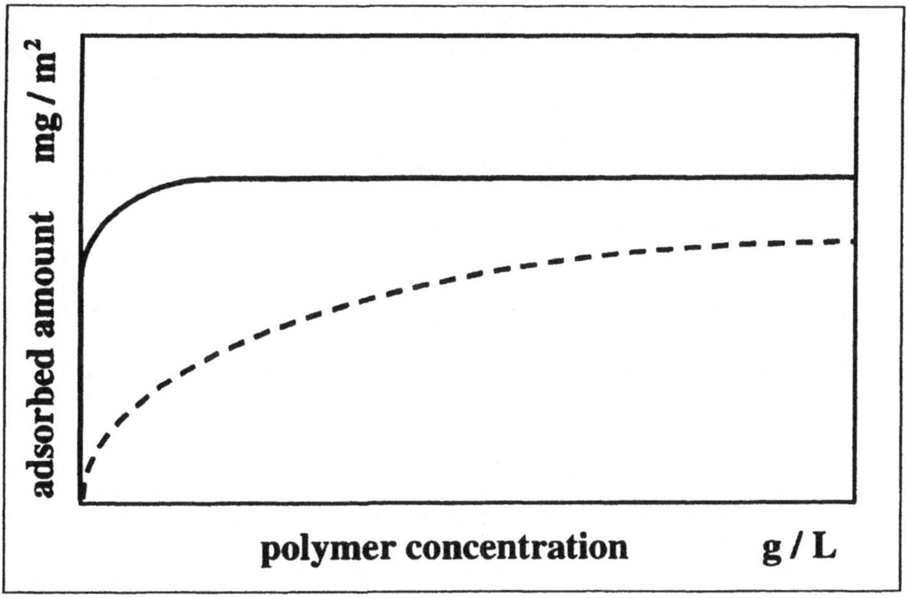

Figure 4.9. Polymer adsorption isotherm.
—— typical adsorption of a monodisperse (i.e., high affinity) polymer. Note that a high degree of adsorption occurs even at low bulk concentration.
------ typical adsorption isotherm of a polydisperse polymer.

For polydisperse polymers usually more rounded isotherms are found together with a less-defined plateau.

Both the molar mass and the solvent quality influence the adsorbed amount. Generally, higher molar masses give higher mass per volume plateau concentrations and adsorbed amounts under theta-conditions are as a rule higher than in good solvents. Theta-conditions are defined as that state of the polymer solution that corresponds thermodynamically to that of an ideal low molar mass solution. The theta-state occurs only at certain temperatures and concentrations in the opposite to real ideal solutions, which behave ideally at all temperatures. Furthermore, the lower the molar mass, the less the high-affinity character of the adsorption isotherm is pronounced. Other factors, which were reported to influence the adsorption, may be the interaction between the solvent and the surface as well as the surface tension of the solvent.[39-41]

The polydispersity-effect has been investigated by several authors.[42-43] The central finding is that, at equilibrium, longer chains adsorb preferentially over shorter ones. However, contrary to a common opinion it is not multiple-anchoring which gives longer molecules an advantage over shorter ones. Adsorption sites on the substrate (stationary phase) are covered with almost equal efficiency by molecules of different length so that the total adsorption energy per unit area in the system is not strongly affected by polydispersity. The entropy of mixing in the solution, however, decreases strongly with increasing chain length and this favors preferential adsorption of longer molecules. Consequently, the driving force for this preference depends on the concentration in solution and is largest for very dilute solutions. The thermodynamic and experimental consequences of polydispersity have been elaborated in a few keynote papers.[43,44]

Another interesting case for displacement chromatography in particular is competitive adsorption of polymers and small molecules. It is often believed that polymers adsorb very strongly and that they therefore tend to displace small molecules or, vice versa, that polymers cannot easily be displaced by small molecules. However this is not necessarily true. Small molecules, which attach more strongly than the segments of the polymer can affect the energy balance and cause polymer desorption (displacement), provided their concentration in solution is sufficiently high. Analytical approximations for both the special case of displacement of polymers by monomeric competitors and polymeric displacers are reviewed by Fleer et al.[45]

Whenever two different polymers compete for an interfacial site, a different situation arises since the translational entropy for both species is so small that it plays almost no role. The situation of lowest free energy is then always the one where the more strongly adsorbing polymer covers the surface. If the more weakly adsorbing polymer is adsorbed first, exchange should eventually occur, although this may be very slow, if kinetic factors enter. A complete description of the kinetics of the polymer adsorption process requires knowledge of the kinetics of all participating elementary processes, including the rates of mass transfer in the bulk solution, the kinetics of the "sticking" process, the kinetics of all conformation changes, and of their dependence on the molar mass, the solvent type and the concentration. These aspect have been discussed elsewhere[45] and will not be elaborated on herein.

The adsorption of polyelectrolytes is influenced by certain parameters in addition to those that govern the adsorption of neutral polymers. Amongst these additional parameters the charge and the ionic strength are the most important. The properties of the substrate (charged or noncharged, charge density) render the process more complicated. This may very well contribute to the fact that experimental findings are often contradictory. At low ionic strength it has been found that highly charged polyelectrolytes adsorb in very small amounts.[46] Under such conditions the electrostatic repulsion along the polymer backbone dominates, resulting in highly extended chains. A very flat adsorption occurs on the surface, which is characterized by a lower mass per charge adsorbed amount than for weakly charged polyelectrolytes. As the ionic strength is increased by addition of salt, the adsorption increases in the majority of cases. Weakly charged polyelectrolytes tend to give higher adsorbed amounts at low ionic strength, however, both upward and downward trends with increasing ionic strength are reported.[47]

A comprehensive overview of polymer adsorption theory and adsorption models are given by Fleer G et al.[45] We will select and discuss only some special aspects which may be helpful in order to explain the experimental findings from displacement chromatography involving oppositely charged stationary phases and polyelectrolytes. However, other combinations including adsorption of polyelectrolytes on uncharged surfaces and the interaction with surfaces bearing the same charge have also been investigated.

In addition to the interacting parameters governing the adsorption of uncharged polymers (adsorption energy parameter and the solvency parameter), the Coulombic interaction is an important aspect of polyelectrolyte adsorption. The ionic forces depend on three main factors:

1. the surface charge/charge density,
2. the polymer charge/charge density,
3. the ionic strength.

Again the pH and the ionic strength may influence the first two of these. Different situations may occur depending on the system parameters:

1. At low ionic strength the electrostatics dominate. Charge compensation is reached if the sum of the surface charge and the polyelectrolyte charge is zero. Effects of molar mass are nearly absent.
2. At high ionic strength the electrostatic contributions are screened. Therefore, the "chemical" interactions become important. Very high adsorption may be observed if the solvent is very poor for the polyelectrolyte under these conditions. At high ionic strength molar mass effects are important.

Figure 4.10 presents schematically the adsorption of a polycation on a negatively charged surface for these two extremes of the ionic strength. For low ionic strength/salt-free solutions (Fig. 4.10a and 4.10b) one always finds charge compensation with a higher degree of adsorption for lower charge densities (Fig. 4.10a) resulting in higher amounts of adsorbed polyelectrolyte than for more highly charged polyelectrolytes. In the case of high ionic strength (Fig. 4.10c and 4.10d) the adsorbed amounts are nearly identical for both low and high charge densities, because the charges are screened. These cases, however, need to be discussed in more detail, i.e., differently for strong and weak polyelectrolytes. In particular, one has to know if pure electrosorption occurs or if the electrosorption is modified by surface affinity. For pure electrosorption it has been found that with increasing salt concentration the ion exchange between small counterions and the polyion causes desorption of the polyelectrolyte. Eventually, all polyions are displaced at a critical salt concentration. There are two models available from which this critical salt concentration (adsorption-desorption transition) for pure electrosorption may be calculated.[48,49] If the electrosorption is enhanced by surface affinity, however, the system becomes less sensitive to the ionic strength.

For strong (fully dissociated), highly charged polyelectrolytes, the desorption found for pure electrosorption at high ionic strength becomes less likely as the surface affinity increases. As a result of the increasing affinity, the polyelectrolyte can compete more effectively with the salt ions, even if the latter are present in excess. For sufficiently high affinity, the adsorbed amount increases in concentrated salt solutions due to the internal screening, which allows some loop and tail formation.[45] The opposite may occur for low charge densities, where the adsorbed amount decreases with increasing ionic strength.[50-53] However, altogether the total adsorbed amount for lower charge densities is expected to be superior to that of higher ones even at high ionic strength.[47]

Whereas for the adsorption of strong polyelectrolytes there is a distinct salt concentration dependence, the ionic strength is much less important for weak polyelectrolytes, which are characterized by a pH-dependent charge density. The reason is that weak polyelectrolytes can adjust their degree of dissociation in order to compensate the surface charge more effectively. As a result, the substrate-adsorbate complex is virtually neutral, which renders the effect of the salt concentration relatively unimportant. Such small ionic strength dependence for weak polyelectrolytes has indeed been found, both theoretically and experimentally.[54,55]

Steric Mass Action Model

The so-called Steric Mass Action (SMA) model, published in 1992 by Brooks and Cramer, constitutes one of the few mathematical tools currently available for the design of displacement separations.[56] Since its first introduction, the model has been fine-tuned and the results have been published in a series of papers.[57-59] To date most of the model's applications are for the simulation of separations of a number of smallish proteins using L- or O-type displacer substances (< 1000 g/mol). The model is based on the following assumptions:

1. The system and all components behave thermodynamically ideal.
2. The multi-point interaction of an adsorbed protein can be expressed as the characteristic charge , ν, of the protein.
3. Adsorption of a large molecule results in the shielding of a number of surface counterions, which no longer are available for ion exchange. The shielded ions may be underneath but also around the adsorbed molecules. In the latter case shielding is due to electrostatic repulsion between the highly charged polyelectrolyte and the small coions. This capacity of the macromolecule is characterized by the steric factor, δ.

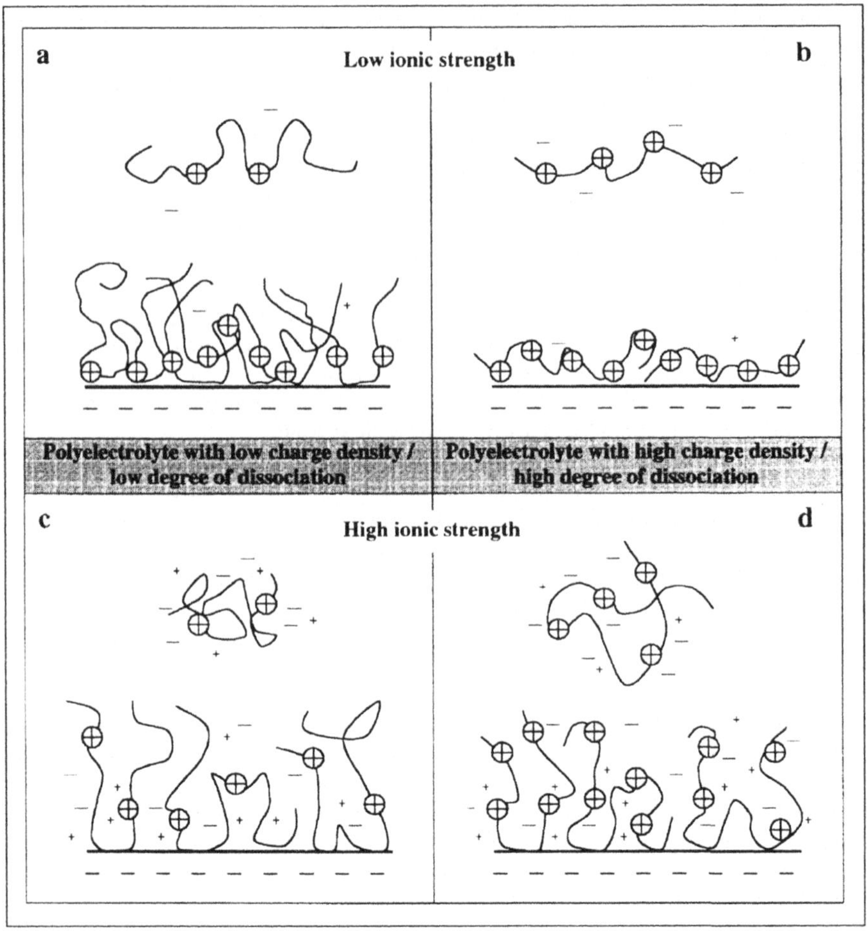

Figure 4.10. Schematic presentation of the adsorption of a polycation on a negatively charged surface for various conditions of charge density and ionic strength.

4. The stationary phase binding affinities (equilibrium constant, K) of each protein and of the displacer are constant and independent of the mobile phase composition (e.g., the salt concentration). Nonideal behavior such as aggregation or changes in the tertiary structure does not occur.

The model is used to describe nonlinear protein (macromolecule) separations by ion exchange chromatography and as such it is not necessarily restricted to displacement chromatography.[60] The model takes into account that salt gradients are induced in front of the displacer zone (caused by the displacement of the small counter ions by the adsorbing displacer) and that therefore each protein finds itself in a specific microenvironment. The three parameters (characteristic charge, steric factor and equilibrium constant) used to describe and predict displacement separations can easily be determined experimentally (Insert 4.2).

The **characteristic charge**, n, of a substance is calculated from isocratic elution data using equation 1:

$$\log k' = \log\left(\phi K \Lambda v\right) - v \log c$$ Eqn. 1

With k′: capacity factor, ϕ: phase ration, K: equilibrium constant, Λ: stationary phase capacity (for monovalent salt counterions), c carrier salt concentration.

Λ is calculated from the breakthrough volume of a 50 mM $NaNO_3$ solution measured at 250 nm.

The **equilibrium constant**, K, can for proteins usually also can be calculated from equation 1. For high affinity substances (e.g., the displacer) K can be calculated from the slope of the linearized single compound isotherm, using the Langmuir formalism. Alternatively, equation 2 may be used after the characteristc charge and the steric factor of the displacer have been determined independently:

$$K = \frac{1}{\phi}\left(\frac{V_B}{V_O} - 1\right)\left(\frac{c_s}{\Lambda - \left(\nu_D + \sigma_D\right)\frac{c_D}{\phi}\left(\frac{V_B}{V_O} - 1\right)}\right)^{\nu_D}$$

Eqn. 2

with V_B: break through volume of the displacer front, V_O: column dead volume and c_s: carrier salt concentration.

The **steric factor**, σ, can be calculated from two points in the non-linear range of the respective substance absorption isotherm according to equation 3.

$$\sigma = \left(\frac{\left(\frac{c_s}{\left(K_i \frac{c_i}{q_i}\right)^{1/\nu_i}}\right) - \Lambda}{-Q_i}\right) - \nu_i$$

Eqn. 3

with c_s: buffer salt concentration, c_I: concentration of the substance under consideration in the mobile phase, and q_I: concentration of the substance under consideration adsorbed to stationary phase (per ml of stationary phase).

Insert 4.3.

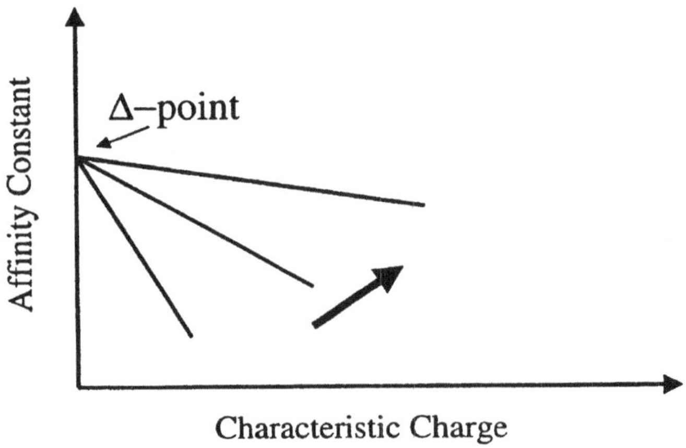

The affinity plot represents the equilibrium constant (log K) as a function on the characteristic charge with the Δ-point as the axial intercept. The Δ-point correspnds to the c_d/q_d ratio of the system. For a chosen displacer concentration Δ, it is simply taken from the displacer isotherm. The affinity plot is created by connecting the Δ-point of the system with the points defined by the log K/ν points of the individual substances. The order in the displacement train will be in the direction of the arrow.

The parameters can subsequently be used to calculate the affinity (see Insert 4.3) and the operating regime plot (Insert 4.4). The affinity plot allows a prediction concerning the probability of a displacement, the order of appearance of the individual substances in the displacement train and the possible consequences of a change in the displacer concentration on a displacement. The operating regime plot defines conditions (displacer and carrier salt concentrations) which still allow a displacement as opposed to those, which would cause an elution ahead of the displacement train in the induced salt gradient. The latter is favored by high displacer concentrations (high induced salt gradients) and high carrier salt concentrations (high salt background). A second line (desorption line) separates conditions enabling displacement from those under which a given substance—instead of being displaced—would be desorbed in the displacer zone.

According to the SMA model, the characteristic charge and the affinity constant together determine the affinity of a given solute in ion exchange chromatography, rather than the affinity constant alone as it was assumed in the conventional approaches. A dynamic affinity parameter, λ, can be defined as follows:

$$\lambda = \sqrt[\nu_i]{K_i/\Delta}$$

with Δ: slope of the operating line.

This approach has been productive in explaining some of the experimental observations, which could not be explained by the assumption of constant separation factors and similar characteristic charges. The SMA approach allows the simultaneous treatment of molecules that differ considerably in size (steric factor) and characteristic charges. Together with a better understanding

Insert 4.4.

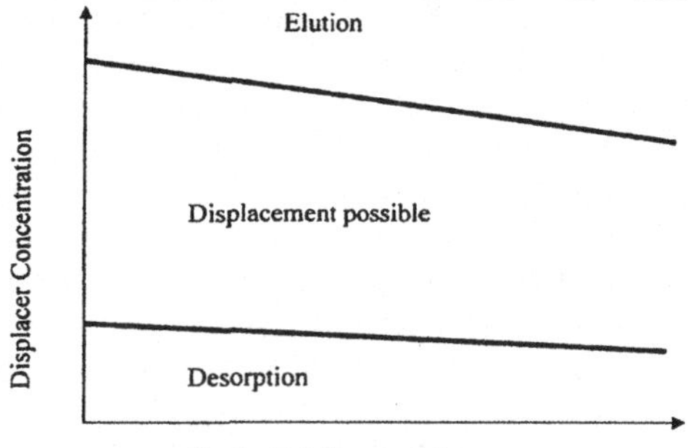

For the operating regime plot the elution line and the displacement line have to be calculated for each substance. The elution line is calculated by an itinerate approach. In the following equation the values for q_d on both sides are changed in such a matter that they become nearly equal.

$$K_D\left(\frac{\Lambda-(\nu_D+\sigma_D)q_D}{c_s}\right)^{\nu_D}=K_i\left(\frac{\Lambda}{c_s+\nu_D\frac{q_D}{K_D}\left(\frac{c_s}{\Lambda-(\sigma_D+\nu_D)q_D}\right)^{\nu_D}}\right)^{\nu_i} \qquad \text{Eqn. 1}$$

With index s: property of the buffer salt, index d: property of the displacer, index I: property of the individual protein.

The q_d value is then used to calculate the c_d value for the respective salt concentration. From a number of such points the elution line ($c_d = f(c_s)$) can be drawn:

$$c_D=\left(\frac{q_d}{K_D}\right)\left(\frac{c_s}{\Lambda-(\nu D+\sigma_D)q_D}\right)^{\nu_D} \qquad \text{Eqn. 2}$$

For the desorption line the Δ-point is calculated as follows:

$$\log\Delta=\frac{v_D\log K_i-v_i\log K_D}{(v_D-v_i)} \qquad \text{Eqn. 3}$$

This Δ-point allows us to calculate the displacer concentration c_d as a function of the salt concentration c_s:

$$c_D=\frac{\Lambda-c_s\left(\frac{\Delta}{K_D}\right)^{v_D}}{(\sigma_D-v_D)\Delta} \qquad \text{Eqn. 4}$$

of how secondary, nonspecific contributions to the affinity influence the affinity, this allows a more rational approach to displacer synthesis.

Systematic Displacer Design, Some Theoretical Considerations

The systematic design of a protein displacer requires a well-developed understanding of the parameters that influence the efficacy of such molecules. Unfortunately we are not yet in a position to fully understand the connection between the physicochemical properties of the displacer molecules and their performance. The database for ion exchange displacement chromatography is comparatively well developed and below three important aspects of displacer performance in such systems will be discussed: the induced salt gradient, the displacer size and chemistry and the stationary phase chemistry as it relates to the displacer chemistry. Some general guidelines can be proposed which should in principle also apply to other types of chromatography.

The sensitivity of the dynamic affinity, λ, to changes in the operating conditions is given by the following equation:[10]

$$d\lambda/\lambda = -\lambda/\nu \, d\Delta/\Delta.$$

According to this equation, L- and O-type displacers with low characteristic charges are more liable to changes in affinity with changing operating conditions than highly charged P-type displacers. The key challenge to systematic displacer design is the creation of a high affinity L- or O-type displacer.

In ion exchange displacement chromatography, as the displacer binds to the stationary phase, small counter ions are liberated (displaced) which form an induced salt gradient ahead of the displacer front.[50] The induced salt gradient depends on the characteristic charge and the concentration of the displacer. As a consequence, each protein in the displacement train finds itself in an individual microenvironment, and it is the salt content of this microenvironment, which determines the protein's isotherm and hence its chromatographic behavior and not the salt content of the bulk mobile phase. The stationary phase affinity of proteins is reduced by an increasing salt concentration of the mobile phase (suppressed/lowered isotherms). This lowering of the protein isotherm by the increased local salt content of the carrier has some important consequences for the displacement:

1. Sufficiently high salt concentrations may lower the isotherm enough to cause the elution of the protein ahead of the displacement train.
2. Low (local) salt concentrations result in relatively higher protein concentrations in the displacement train.
3. Low salt concentrations result in a high stationary phase capacity (better development of the displacement train, which is a function of the stationary phase capacity).
4. Fine-tuning the protein affinity via the (local) salt concentration can reduce tailing and mixing of the individual protein zones.

Most displacement separations in the ion exchange mode are currently carried out at low carrier salt content to improve performance. However, there is a lower limit beyond which desorption kinetics become too slow for practical purposes.

It has also been observed that under otherwise identical conditions, L-type displacers perform better if carriers with medium rather than low salt concentrations are used.[61] In hindsight this can be explained by the reduction of the target proteins stationary phase affinity by the induced salt gradients. P-type ion exchange displacers (polyelectrolytes), on the other hand, work equally well at low salt concentration, since the isotherms of such molecules are almost independent of the local salt environment. As displacers, this renders them less dependent on the process parameters, but also makes column regeneration more difficult.

Size is one of the best-investigated displacer properties in ion exchange displacement chromatography. Since affinity increases with size, large polyelectrolytes have high, sometimes too high, stationary phase affinities. At the same time, they are associated with problems with regard to heterogeneity, solubility, mass transfer, steric exclusion, viscosity increase of the mobile phase, and column regeneration. By comparison, small molecules have lower affinities, but are more suited to the development of a rugged industrial process. By comparison, small molecules have:

1. lower characteristic charges (number of charges per molecule)
2. lower steric factors (number of shielded sites on the stationary phase surface)
3. lower affinity constants
4. lower steric factor to characteristic charge ratios (δ/ν) and hence
5. higher normalized characteristic charge (characteristic charge per repeating unit)

The last point has consequences for the induced salt gradient since for a given number of repeating units, the L-type molecules displace relatively more salt ions with the predictable effects on the salt microenvironments in the displacement train. The physical basis for these observations remains unclear; however, an decrease in the normalized characteristic charge with increasing molecular mass can only be explained by a prevention of the interaction of a certain number of charges in the polyelectrolyte, perhaps for sterical reasons. The result is a less dense surface covering by the P-type displacer and a smaller induced salt gradient.

By manipulating the characteristic charge in connection to the mass of the displacer molecule, one can presumably fine-tune the dynamic affinity of the displacer and the importance of the salt gradient induced by it. In this context a comparison between two aminoglycosidic antibiotics (Neomycin B and Streptomycin A) should be noted.[11] Streptomycin contains two strongly basic guanidine moieties and Neomycin six primary amines. Even though the primary amines constitute weaker individual interaction sites, the characteristic charges were 2.3 for Streptomycin and 4.5 for Neomycin A, respectively. Neomycin A was subsequently found to be a high affinity L-type displacer capably of displacing even lysozyme from a cation exchanger stationary phase. In spite of its high affinity, however, Neomycin could be removed from the column by relatively mild conditions.

Steric factor and characteristic charge are not fixed molecular properties. They depend also on the surface chemistry (charge density and distribution) of the stationary phase and hence can be optimized from that side. Even though the dynamic affinity of the L-type displacers tends to be lower than that of P-type, it is generally considerably higher than that of any protein. Proteins have comparatively high steric factor to characteristic charge ratios due to their high conformational constrains. This explains why L-type displacers, such as protected amino acids (arginine, lysine esters), can displace proteins which use (several) arginine and lysine moieties to interact with the stationary phase.[10]

The surface chemistry of the intended stationary phase is an important aspect of displacer design. It is commonly assumed that existing stationary phases for elution chromatography are equally suited for displacement chromatography. This is not necessarily the case. A displacer in ion exchange chromatography will be a polyelectrolyte. The many parameters that can influence the adsorption of such molecules have been discussed previously. While specific electrostatic interactions with the stationary phase are most important for the binding of such molecules to the stationary phase, secondary interactions (e.g., hydrophobic, π-π) are also possible. A recent comparison of the performance of various O-type displacers on three typical types of ion exchangers, namely a poly(methacrylate)-based material (PMA), a hydrophilized poly(styrene-divinylbenzene)-based material (PS-DVB) and an agarose-based material gave the following indications.[6,62]

1. The affinity of the displacer increases with the number of charges.
2. Up to a certain point, the affinity also increases with the spacing of the charges within the displacer molecule.
3. Generally, linear and more flexible molecules have higher affinities than branched or cyclic ones with equal charge densities.
4. Secondary interactions may strongly influence the affinity.
5. The positioning of the additional interaction sites within the displacer molecules is important. Aromatic functions, for example, are more effective in increasing the affinity when placed at the outer edge rather than in the center of the ionic displacer.

The last two points offer interesting options for the design of a high-affinity L-type displacer. In this context the stationary phase is of utmost importance. In the case of the above mentioned three types of ion exchangers it was found, for instance, that the introduction of a hydrophobic region increased the displacer affinity only in the case of the PMA and the PS-DVB materials. Aromatic groups were only effective in the case of the PMA based column. In the case of agarose, no affinity increase was observed, which is not surprising given the highly hydrophilic nature of this material.

Conclusions

Displacement chromatography has an enormous potential as a preparative bioseparation technique. In many situations from the mg to the kg scale and beyond the displacement chromatography may theoretically be the most practical, the most economic and the most efficient approach to a given separation problem. However, in order to exploit the full potential of displacement chromatography, suitable displacer/stationary phase systems must become available. This chapter is intended as an introduction to our current understanding of the requirements for systematic displacer design.

References

1. Guiochon G, Golsham-Shirazi S, Katti A. Fundamentals of preparative and nonlinear chromatography. San Diego: Academic Press, 1994.
2. Freitag R. Displacement chromatography—Application to downstream processing in biotechnology. In: Subramanian G, ed. Bioseparation and Bioprocessing. Vol 1 Biochromatography. Weinheim: VCH Verlagsgruppe, 1998:89-112.
3. Gerstner JA. Economics of displacement chromatography—A case study: Purification of oligonucleotides. Biopharm 1996; 9:30-36.
4. Kasper C, Vogt S, Breier J et al. Protein displacement chromatographyin hydroxy- and fluoroapatite columns. Bioseparation 1996; 6:247-262.
5. Frenz J, Bourell J, Hancock WS. High-performance displacement chromatography-mass spectrometry of tryptic peptides of recombinant human growth hormone. J Chromatogr 1990; 512:299-314.
6. Shukla AA, Barnthouse KA, Bae SS et al. Structural characteristics of low-molecular-mass displacers for cation-exchange chromatography. J Chromatogr 1998; 814:83-95.
7. Cramer S, Subramanian G. Recent advances in the theory and practice of displacement chromatography. Sep Purif Meth 1990; 19(1):31-91.
8. Freitag R. Displacement chromatography of biomolecules. In: Aboul-Einein HY, Ed. Analytical and Preparative Separation Methods for Biomolecules. Marcel Dekker, 1999:203-252.
9. David Jen SC, Pinto NG. Use of the sodium salt of poly(vinylsulfonic acid) as a low-molecular-weight displacer for protein separations by ion-exchange displacement chromatography. J Chromatogr 1990; 519:87-98.
10. Kundu A, Vunnum S, Jayaraman G et al. Protected amino acids as novel low-molecular-weight displacers in cation-exchange. Biotechnol Bioeng 1995; 48:452-460.
11. Kundu A, Vunnum S, Cramer SM. Antibiotics as low-molecular-mass displacers in ion-exchange displacement chromatography. J Chromatogr 1995; 707:57-67.

12. Freitag R, Vogt S. Preparation of plasmid DNA using displacement chromatography. Cytotechnol 1999; 30:159-167.
13. Antia FD, Fellegvari I, Horvath C. Displacement of proteins in hydrophobic interaction chromatography. Ind Eng Chem Res 1995; 34:2796-2804.
14. Ruaan RC, Hsu D, Chen WY et al. Protein separation by hydrophobic interaction chromatography using methacrylic block copolymers as displacers. J Chromatogr 1998; 824:35-43.
15. Zhu J, Katti AM, Guiochon G. Effect of displacer impurities on chromatographic profiles obtained in displacement chromatography. Anal Chem 1991; 63:2183-2188.
16. David Jen SC, Pinto NG. Dextran sulfate as a displacer for the displacement chromatographyof pharmaceutical proteins. J Chromatogr Science 1991; 29:478-484.
17. Galaev IY, Arvidsson P, Mattiasson B. Polymer displacement in dye-affinity chromatography. J Chromatogr 1995; 710:259-266.
18. Vogt S., Freitag R. Comparison of anion exchange and hydroxyapatite displacement chromatography for the isolation of whey proteins. J Chromatogr 1997; 760:125-137.
19. Peterson EA. Ion-exchange displacement chromatographyof serum proteins, using carboxymethyldeytranes as displacers. Anal Biochem 1978; 90:767-784.
20. Torres AR, Edberg SC, Peterson EA. Preparative high-performance liquid chromatography of proteins on an anion exchanger using unfractionated carboxymethyl dextran displacers. J Chromatogr 1987; 389:177-182.
21. Torres AR, Peterson EA. Purification of complex protein mixtures by ion-exchange displacement chromatographyusing spacer displacers. J Chromatogr 1992; 604:39-46.
22. Peterson EA, Torres AR. Ion-exchange displacement chromatographyof proteins, using narrow-range carboxymethyldextrans and a new index of affinity. Anal Biochem 1983; 130:271-282.
23. Torres AR, Dunn BE, Edberg SC. Preparative high-performance liquid chromatography of proteins on ion-exchange columns with carboxymethyldextrans as displacers. J Chromatogr 1984; 316:125-132.
24. Torres AR, Peterson EA. Purification of complex protein mixtures by ion-exchange displacement chromatographyusing spacer displacers. J Chromatogr 1992; 604:39-46.
25. Freitag R, Breier J. Displacement chromatographyin biotechnological downstream processing. J Chromatogr 1995; 691:101-112.
26. Freitag R, Vogt S, Mödler M. Thermoreactive displacers for anion exchange and hydroxyapatite displacement chromatography. Biotechnol Progr 1999; 15(3):573-576.
27. Patrickios CS, Gadam SD, Cramer SM et al. Block methacrylic polyampholytes as protein displacers in ion-exchange chromatography. Biotechnol Prog 1995; 11:33-38.
28. Quintero G, Vo M, Farkas G et al. Series of homologous displacers for preparative chiral displacement chromatographic separations on cyclobond-II columns. J Chromatogr 1995; 693:1-5.
29. Stigter D. The charged colloidal cylinder with a Gouy double layer. J Colloid Interface Sci 1975; 53:296-306.
30. Manning GS. The molecular theory of polyelectrolyte solutions with applications to the electrostatic properties of polynucleotides. Quart Rev Biophys 1978; 11:179-246.
31. Fixman M. The Poisson-Boltzmann equation and its application to polyelectrolytes. J Chem Phys.1979; 70:4995-5005.
32. Odijk T. On the limiting law solution of the cylindrical Poisson-Boltzmann equation for polyelectrolytes. Chem Phys Lett 1983; 100:145-150.
33. Manning GS, Zimm BH. Cluster theory of polyelectrolyte solutions. I Activity coefficient of the mobile ions. J Chem Phys 1965; 43:4250-4259.
34. Manning GS. Cluster theory of polyelectrolyte solutions. II. Additivity rules, osmotic pressure, and Donnan equilibrium. J Chem Phys 1965; 43:4260-4267.
35. Manning GS. Limiting laws and counterion condensation in polyelectrolyte solutions. I. Colligative properties. J Chem Phys 1969; 51:924-933.
36. Dautzenberg H, Jaeger W, Kötz J, Philipp B et al. Polyelectrolytes: Formation, Characterization, Application. München: Carl Hanser Verlag, 1994:249.
37. Zeitz K. Zur Löslichkeit eines Kationischen Polyelektrolyten in Wässrigen Salzlösungen, PhD Thesis: TU Berlin, 1995.
38. Wandrey C. Concentration regimes in polyelectrolyte solutions. Langmuir. 1999; 15:4069-4075.

39. Cohen Stuart MA, Fleer GJ, Bijsterbosch BH. The adsorption of poly(vinyl pyrrolidone) onto silica. I. Adsorbed amount. J Colloid Interface Sci 1982; 90:310-320.
40. Griot O, Kitchener J. A. Role of surface silanol groups in the flocculation of silica suspensions by polyacrylamide. Part 1.- Chemistry of the adsorption process. Trans Faraday Soc 1965; 6:1026-1031.
41. Ober R,. Paz L, Taupin C et al. Study of surface tension of polymer solutions: Theory and experiments. I. Good solvent conditions. Macromolecules. 1983; 16:50-55.
42. Felter RE, Ray LN. Polymer adsorption studies at the solid-liquid interface using gel permeation chromatography. I. Molecular weight distribution along the adsorption isotherm. J Colloid Interface Sci 1970; 32:349-360.
43. Roe RJ. Selective adsorption of polymers from solution. In: Lee LH, ed. Adhesion and Adsorption of Polymers. New York: Plenum Press, 1980:629-641.
44. Scheutjens JMHM, Fleer GJ. Some implications of recent polymer adsorption theory. In: Tadros TF, ed. The Effect of Polymers on Dispersion Properties. London: Academic Press, 1982:145-168.
45. Fleer GJ, Cohen Stuart MA, Scheutjens JMHM et al. Polymers at Interfaces. London: Chapman & Hall, 1993.
46. Cohen Stuart MA, Fleer GJ, Lyklema J et al. Adsorption of ions, polyelectrolytes and proteins. Adv Colloid Interface Sci 1991; 34:477-535.
47. Durand G, Lafuma F, Audebert R. Adsorption of cationic polyelectrolytes at clay-colloid interface in dilute aqueous suspension—Effect of the ionic strength of the medium. Progr Colloid Polym. Sci 1988; 76:278-282.
48. Wiegel FW. Adsorption of a macromolecule to a charged surface. J Phys 1977; A10:299-303.
49. Muthukumar M. Adsorption of polyelectrolyte chain to a charged surface. J Chem Phys 1986; 86:7230-7235.
50. Froehling RE, Bantjes A, Kolar Z. Adsorption of a synthetic heparinoid polyelectrolyte on an ion-exchanging surface. J Colloid Interface Sci 1977; 62:35-39.
51. Davis RJ, Dix LR, Toprakcioglu C. Adsorption of poly-L-lysine to mica powder. J Colloid Interface Sci 1989; 129:145-152.
52. Hendrickson ER, Neuman RD. Polyacrylamide adsorption from very dilute solutions. J Colloid Interface Sci 1986; 110:243-251.
53. Pelton RH. Electrolyte effects in the adsorption and desorption of a cationic polyacrylamide on cellulose fibers. J Colloid Interface Sci 1986; 111:475-485.
54. Böhmer MR, Evers OA, Scheutjens JMHM. Weak polyelectrolytes between two surfaces: Adsorption and stabilization. Macromolecules 1990; 23:2288-2301.
55. Blaakmeer J, Böhmer MR, Cohen Stuart MA et al. Adsorption of weak polyelectrolytes on highly charged surfaces. Poly(acrylic acid) on polystyrene latex with strong cationic groups. Macromolecules 1990; 23:2301-2309.
56. Freitag R, Horvath C. Chromatography in the downstream processing of biotechnological products Adv Biochem Eng Biotechnol 1995; 53:17-59.
57. Gallant SR, Kundu A, Cramer SM. Modeling nonlinear elution of proteins in ion-exchange chromatography. J Chromatogr 1995; 702:125-142.
58. Gallant SR, Cramer SM. Productivity and operating regimes in protein chromatography using low-molecular-mass displacers. J Chromatogr 1997; 771:9-22.
59. Kundu A, Barnthouse A, Cramer SM. Selective displacement chromatographyof proteins. Biotechnol Bioeng 1997; 56(2):119-129.
60. Iyer H, Tapper S, Lester P et al. Use of the steric mass action model in ion-exchange chromatographic process development. J Chromatogr 1999; 832:1-9.
61. Gadam SD, Cramer SM. Salt effects in anion exchange displacement chromatography, comparison of pentosan polysulfate and dextran sulfate displacers. Chromatographia 1994; 39(7/8):409-418.
62. Shukla AA, Bae SS, Moore JA et al. Structural characteristics of low molecular mass displacers for cation-exchange chromatography, II role of the stationary phase. J Chromatogr 1998; 827:310.
63. Kundu A, Vunnum S, Cramer SM. Displacement chromatographyof proteins using low molecular weight anionic displacers. Adsorption 1998; 4(3,4):373-381.
64. Gerstner JA, Cramer SM. Heparin as a nontoxic displacer for anion-exchange protein displacement systems. BioPharm 1992; 11:42-45.
65. DeVauldt D. The theory of chromatography. J Am Chem Soc 194.

CHAPTER 5

Membrane Adsorbers for Decontamination and Leukocyte Removal

Galya Tishchenko and Miroslav Bleha

Introduction

Membranes were originally developed as agents (filters, sieves) for solid-liquid separations by mechanical means. In this context any interaction (adsorption) of the components with the membrane material was to be avoided. Recently however, the controlled (selective) adsorption of certain molecules on such membranes has become the object of intensive study.[1-9] This does not mean that the primary sieving function of the membrane has lost its importance in such cases. Quite the opposite, it is the combination of the two functions of the membranes, sieving and adsorption, which opens up new possibilities for improving the efficiency of separation processes, especially in areas such as biotechnology and medicine (bioseparation).

Adsorption, on the other hand, has also been used extensively for (bio)separation, often under the name of chromatography. Traditionally such separations by adsorption were carried out using packed beds of beaded adsorbents (columns). Compared to chromatographic columns, membrane adsorbers show considerably faster mass transfer kinetics, and thus can be operated at higher flow-rates and with better throughput.[10-17] The result is a significant shortening of the process time when membrane adsorbers are used, which in addition often causes an improvement of the bioproduct quality and the efficiency of separation processes. Some authors argue that in certain cases and particularly for applications in biotechnology and medicine, membrane adsorbers may in future even supersede beaded supports.

Several types of processes exploiting the adsorptive ability of membranes are intensively developed at present, namely "membrane chromatography",[18-31] "decontamination (mainly depyrogenation)"[32-41] and "removal of leukocytes from blood products."[42-47] The more medical applications of membrane adsorbers include, for example, their use in an adsorbing artificial kidney. In this case a system composed of an ultrafilter (UF) and adsorbent is used, which in addition shows a pronounced tendency for the selective adsorption of uric acid and creatinine from human plasma. The selectivity of such a system can be much higher than, for example, that found for activated carbon.[48] Depyrogenation, preparation of leukocyte-depleted blood products and the adsorbing artificial kidney are typical examples of the application of membranes as adsorbers rather than filters, since the separation relies less on differences in size between the molecules but rather on differences in surface binding strength (affinity). Concomitantly, these applications are different from conventional (membrane or column) chromatography in the sense that the desorption of the sorbed components ("contaminants") is done only for membrane regeneration and sanitation and some membrane adsorbers are even intended for single use only.

Synthetic Polymers for Biotechnology and Medicine, edited by Ruth Freitag.

Membrane chromatography and membrane adsorption may operate hand in hand in the preparation of biopharmaceuticals in particular those intended for human treatment, since the production of biologically active substances from a complex media always involves both purification and decontamination (e.g., depyrogenation or removal of DNA/viruses). Despite the difference in the pursued aim, the main requirement for the membrane in question is the same in both processes, i.e., achievement of the maximum sorption selectivity of the target substance. The efficiency of both processes is affected by the same parameters, which, however, have somewhat different accents in membrane chromatography and decontamination. Primarily, it is the chemical nature of the membrane material (base polymer, type and concentration of functional groups/ligands) together with its morphology (porosity, pore size distribution), which defines the adsorption capacity and the kinetic properties of the membrane. Once an interactive membrane has been chosen, the operation conditions (flow rate, type and concentration of buffer/salt solutions and their pH, sample volume and concentration) must be optimized for the intended process. Especially in the case of a "simple" clean-up step, such as a depyrogenation or leukocyte removal, it will often be impossible to change the pH and the salt concentration of the solutions to be subjected to this step. In such cases, the choice of the membrane, its chemical nature and its morphology, will be the most important variable during process development. Fortunately, a wide variety of membrane materials (hydrophilic, hydrophobic, positively charged, pseudoaffinity) based on various polymers (cellulose acetate, nitrocellulose, poly(vinyl acetate), nylon, polycarbonate, poly(ethylene terephthalate), polyester, polysulfone, polyethersulfone, poly(vinylidene difluoride), polyethylene, polypropylene and polymer alloys) are now available for these purposes. Table 5.1 gives an overview of the more important structures.

In this chapter we attempt to find a correlation between the chemical nature of the membrane and its suitability to typical clean-up steps such as depyrogenation and leukocyte removal from blood-products. The experimental results of many investigators are compared in this context in regard to the time-average adsorption capacity and the time-average removal factor. The dependence of these two process parameters on the average pore size and the chemistry of the adsorptive membrane are also analyzed.

Evaluation of the Adsorbing Efficiency of Interactive Membranes

The adsorption capacity (A) is usually determined using equation 1.

$$A = (C_f - C_p) \times V / S \ [mg/m^2] \qquad (1)$$

with: C_f, C_p concentration of the substance of interest in the feed and the permeate [mg/ml], respectively; V volume of the feed solution [mL]; S filtration area of membrane adsorber [m^2].

Any comparison of adsorptive membranes in relation to their adsorption capacity loses significance if the operation conditions (initial solute concentration, volume of feed solution, flow rate, time of filtration) vary largely between the different experiments. In this case, i.e., for an "absolute" evaluation of the efficiency of a given adsorptive membrane, we propose to use the time-average adsorption capacity (A_T) instead, i.e., the adsorption capacity per unit of time (t), equation 2:

$$A_T = (C_f - C_p) \times V / S\,t \ [mg/m^2h] \qquad (2)$$

The time-average removal factor (F_T) can be calculated as follows, equation 3.

$$F_T = (C_f / C_p) \times V / S\,t \ [m/h] \qquad (3)$$

In contrast to the removal factor (F), which is usually used for evaluation of the process efficiency, the time-average removal factor (F_T) contains an additional term which takes into

Table 5.1. Synthetic membrane materials

Polymer		Structure
Polyamide (Nylon)	PA	$-[NH-(CH_2)_x-NHC(=O)-(CH_2)_y-C(=O)]_n-$
Copolymer of ethylene and vinyl alcohol	PEVA	$-[(CH_2-CH_2)_x(CH_2-CH(OH))_y]_n-$
Cellulose ester (Hydroxyethyl)cellulose Cellulose acetate Nitrocellulose (Diethylaminoethyl) cellulose	CE HEC CA NC DEAE	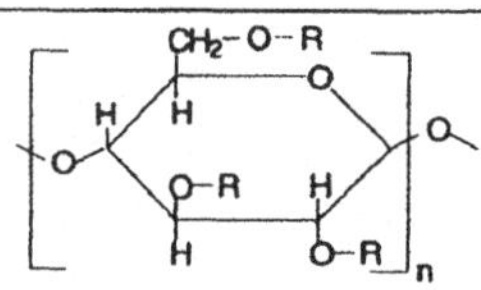 R - COalkyl, CH_2CH_2OH, $COCH_3$, NO_2, $CH_2CH_2N(C_2H_5)_2$
Polyethylene Poly(methyl methacrylate) Poly(acrylonitrile) Poly(vinyl acetate)	PE PMMA PAN PVA	$-(CH_2-CR_1R)_n-$ R-H, $COOCH_3$, CN, R_1-H PVA: R-$COOCH_3$, R_1-CH
Poly(ethylene terephthalate)	PET	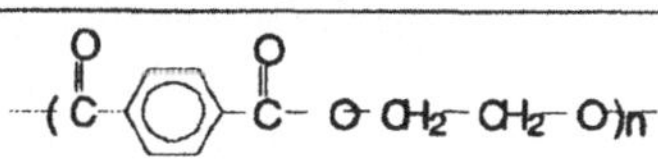
Polyurethane	PU	$-(C(=O)-NH-X-NH-C(=O)-O-Z-O)_n-$
Polyester	PES	$-(C(=O)-Z-C(=O)-O-X-O)_n-$
Poly(tetrafluoroethylene)	PTFE	$-(CF_2-CF_2)_n-$
Polycarbonate	PC	$-(C(=O)-O-Z-O)_n-$ X, Z = alkanediyl or arenediyl
Poly(vinylidene difluoride)	PVDF	$-(CH_2-CF_2)_n-$
Polysulfone Polyethersulfone	PS PESF	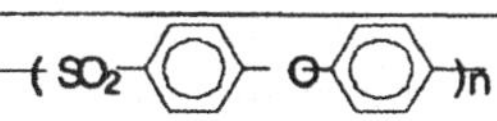

account the volume of the feed solution passed through the filtration area unit (1 m^2) per time unit (1 h). The F_T value depends thus on the ratio between the time-average amount of solute passed through the membrane (C_f x V/S t) and the final target molecule concentration in the permeate. In addition, the F_T value depends on the rate of change in the C_f/C_p-ratio under the applied conditions (linear flow rate of the feed solution, V/S t).

Membranes for Depyrogenation (Endotoxin Removal)

The terms endotoxin, pyrogen and lipopolysaccharide (LPS) are used almost synonymously for a certain class of pyrogenic (i.e., fever-inducing) substance. Such endotoxins in nature are found in the outer membranes of gram-negative bacteria (for example *E. coli*) and blue-green algae (cyanobacteria).[49] These microorganisms are almost ubiquitous in nature and can grow even in distilled water. During bacterial growth and after cell death, their LPS may be released into the environment. LPS is known to cause violent reactions in mammals with typical symptoms of high fever, vasodilation, diarrhea and, in extreme cases, death by septic shock. Fever is probably one of the least dangerous of the numerous effects of the bacterial LPS. Next to pulmonary and cardiodepressant effects, major influences on the immune system, such as activation of the B-lymphocytes and mononuclear cells, and stimulation of the release of interleukin-1, interferons, prostaglandins, leukotrienes and tumor-necrosis factor have been ascribed to administration of LPS.[49]

In humans, contact with LPS may occur not only during a bacterial infection but also via LPS-contaminated medicaments and solutions administrated intravenously (parenteralia). Since the biological effects of LPS may appear even at concentrations of 1 ng per 1 kg of body weight, drugs intended for parenteral use, have to be "endotoxin-free", i.e., thoroughly depyrogenated. In fact parenteralia have to comply with LPS threshold limits (in EU[1], endotoxin unit) regulated by pharmacopoeias. For example, tetracycline hydrochloride may not contain more than 0.5 EU/mg. Similar limits exist for insulin (0.8 EU/insulin unit), hyaluronidase (2.3 EU/ hyaluronidase unit), the sodium salt of heparin for injection (0.003 EU/heparin unit).[50]

Clearly, every conceivable precaution has to be taken in large-scale production of parenteralia to avoid pyrogen contamination of the final product. Contamination with LPS may occur at a number of sites and may originate from many different sources, such as the raw materials (water, salts, additives), the production organism (such as *E. coli*) or the chromatographic column (stationary phase). The situation is especially critical under conditions where up to date technology is not available. Examples of such situations are local pharmacies and hospitals in developing countries.

The removal of endotoxins is difficult. Chemical treatment with acids, alkalis or oxidizing agents and/or autoclaving are commonly practiced methods for sterilization. However, these are not suitable for the preparation of LPS-free solutions. A general approach to reduce or avoid LPS contamination is the use of LPS-free raw materials, along with LPS-free equipment and short processing/packaging procedures, which avoid contamination and microbial growth during production and filling. However, under certain conditions, LPS-free raw materials may not be available (e.g., tissue culture media, monoclonal antibodies and radiopharmaceuticals, the production of which is extremely complex), or their quality may be questionable (water for preparation of hemodialysis solutions). Moreover, if recombinant DNA methods are exploited, introduction of host-specific LPS, accidental microbial contamination during downstream processing or copurification of LPS with the product may also take place.[52]

Depyrogenation thus quickly becomes a critical point in the formulation especially of high-molecular-mass pharmaceuticals such as proteins. Amongst the many methods for depyrogenation, membrane adsorbers appear to have a number of important advantages. Their application for endotoxin removal as well as aspects to consider during process development is therefore discussed below.

Physicochemical Properties of LPS

The efficiency of a clean-up step by (membrane) adsorption depends, in addition to the chemistry of the membrane, strongly on the physicochemical properties of the substance to be adsorbed. The physical state of LPS in aqueous media depends among other things on the bacterial strain from which it was derived and composition of the suspending liquid. LPS derived from wild-type bacteria are classified into three categories in terms of their molecular structure.[51] Type I LPS is composed of a so-called O-antigen-specific polysaccharide, the outer- and inner-core polysaccharides, and lipid A (Fig. 5.1). Type I endotoxin is typically found in *Enterobacteriaceae* and *Pseudomonas* species. Type II LPS lacks the O-antigen-specific polysaccharide and is found, for example, in *Hemophilus influenzae*, *Acinetobacters*, and *Bordetella pertussis*. Type III LPS consists of an inner core oligosaccharide and lipid A and is found in *Chlamydia* species.

In spite of the difference in molecular composition, the chemical properties of various types of LPS are closely related due to the consistent presence of phosphate groups and lipid A. The LPS monomer is amphiphilic since it contains both a large hydrophilic polysaccharide chain and a strongly hydrophobic fatty acid tail. The sugar residues may be partially phosphorylated. The pyrogenic (fever inducing) properties of the LPS are generally associated with the lipid A part of the structure, shown in detail in Figure 5.2.[46] In its native state, lipid A anchors the LPS-molecule to the outer membrane of the bacterium. The pK values of the primary phosphate group of lipid X, a pyrogen precursor, were determined as $pK_1 = 1.3$ and $pK_2 = 8.2$. At pH 1.3, the phosphate groups in the LPS molecule are completely ionized and the molecule carries a negative charge. In water, blood (near neutral pH) and aqueous solutions in general, the LPS molecule also bears a negative charge.

Monomeric LPS has a molecular mass in the order of 10,000-20,000. The size of the monomeric form of type II LPS is much smaller than the size of type I LPS and thus type II is less easily removed from solution by sieving. In addition, LPS molecules show a pronounced tendency for aggregation in aqueous solution.[53-55] Type I LPS aggregates tightly and is not even dissociated by heat exposure, whereas the aggregates of type II LPS are heat labile. However, at least in an aqueous environment, LPS can be considered to have a molecular mass of 10^6, corresponding to a size in the angstrom-range. In the presence of divalent cations, such as Ca^{2+}, LPS molecules may form aggregates with a diameter of 100 nm and molecular masses of up to 10^8.[56]

Important conclusions can be drawn from this analysis of the physicochemical properties of LPS in regard to any depyrogenation of contaminated liquids. The pharmaceutical industry usually recommends a so-called "sterile filtration", in which the liquid is first passed through a prefilter with pore sizes > 0.2 mm (often 0.45 mm) and subsequently through an UF filter with an average pore size of 0.2 mm. This may decrease the level of product contamination but does not necessarily guarantee an LPS-free product. The prefilter is expected to retain particulate contaminants, while the UF filter should prevent penetration of LPS aggregates into the filtrate. However, the experimental results concerning residual LPS levels in the filtrate are confusing and inconsistent. Below, the use of membrane-based separations for depyrogenation are discussed in more detail, taking not only sieving but also the ever-present possibility of adsorption into account.

Membranes Based on Nonfunctionalized Polymers for Endotoxin Removal

In contrast to membrane filtration, in which adsorption is an undesirable phenomenon because of the possible loss of target substance, membrane adsorption is in fact based on this phenomenon and high sorption capacities improve in this case the efficiency of the operation from the economical point of view. However, virtually all membrane materials (surfaces), even the ones designed primarily with membrane filtration in mind, exhibit a more or less pronounced but nonspecific adsorption tendency, due to the general possibility of establishing hydrophobic, dipole-dipole and electrostatic interactions as well as hydrogen bonds with the various dissolved molecules. It was also observed by many investigators that mechanical filter may retain particles/molecules that are smaller than the filter's pores, presumably as a result of an impaction of the particles on the filter surface. However, whatever the cause, if the forces

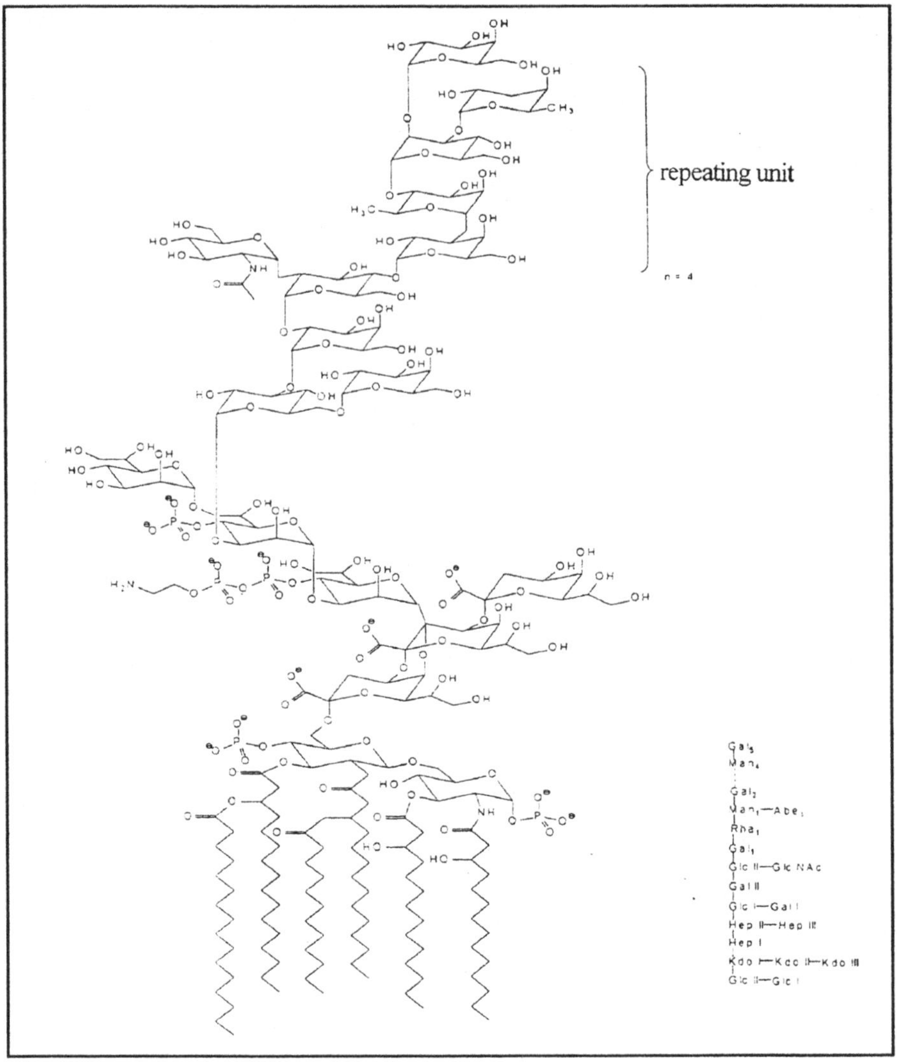

Figure 5.1. Model of a generic LPS molecule as proposed by Kastowsky et al. The molecule is composed of the lipid A structure from *E.coli*, the oligosaccharide region from *S. typhimurium* and the O-specific side chain from Salmonella serogroup B.

holding the particles/molecules to the filter are not strong enough, these molecules will sooner or later detach from the filter and pass it into the filtrate.

The possibility of aggregation and adsorption of the LPS molecules may have added to the inconsistent results obtained for residual LPS levels in the filtrate of sterile filtrations, as summed up in Table 5.2. The filter membranes used in these experiments generally do not intentionally bear "interactive" groups. Nevertheless, surprisingly high LPS adsorption capacities were observed in certain cases. In fact, in a number of cases, where a nonfunctionalized membrane has been used successfully for depyrogenation, of mostly salt solutions for hemodialysis, saccharide solutions and water (distilled, tap, subterranean),[46,49,50,56,57] the separation by sieving

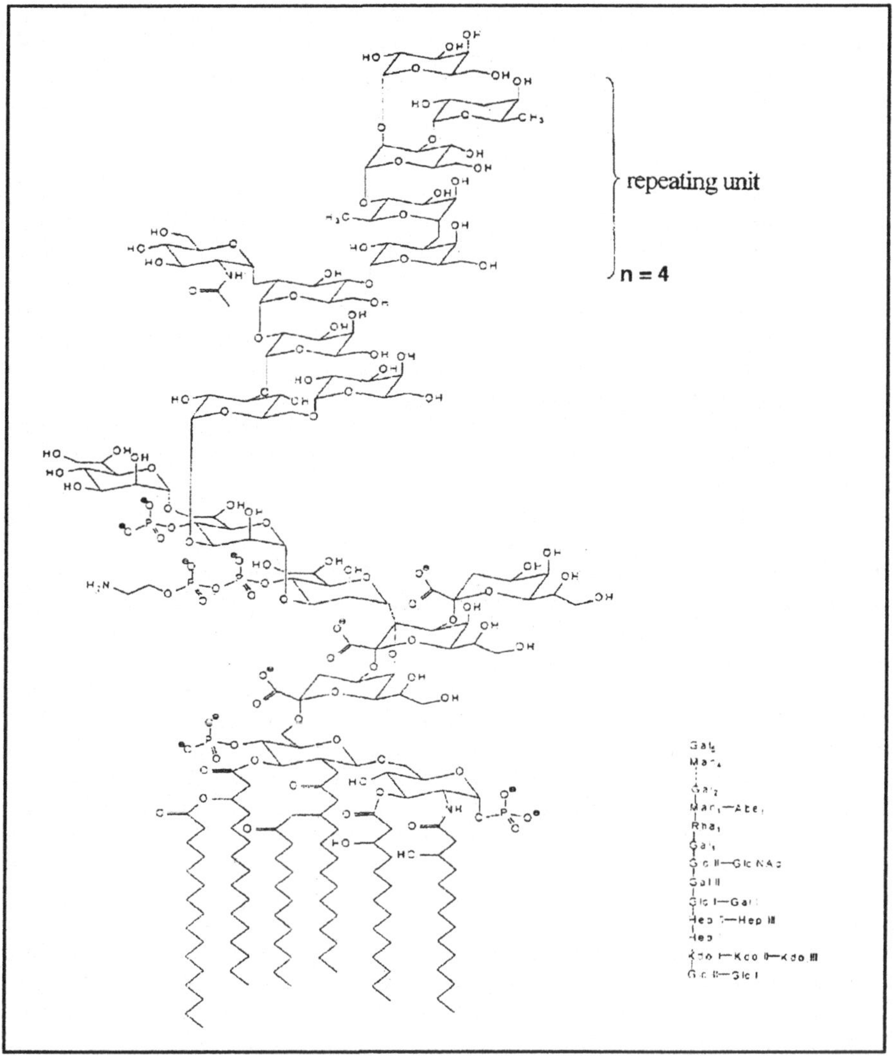

Figure 5.1. Model of a generic LPS molecule as proposed by Kastowsky et al. The molecule is composed of the lipid A structure from *E.coli*, the oligosaccharide region from *S. typhimurium* and the O-specific side chain from Salmonella serogroup B.

holding the particles/molecules to the filter are not strong enough, these molecules will sooner or later detach from the filter and pass it into the filtrate.

The possibility of aggregation and adsorption of the LPS molecules may have added to the inconsistent results obtained for residual LPS levels in the filtrate of sterile filtrations, as summed up in Table 5.2. The filter membranes used in these experiments generally do not intentionally bear "interactive" groups. Nevertheless, surprisingly high LPS adsorption capacities were observed in certain cases. In fact, in a number of cases, where a nonfunctionalized membrane has been used successfully for depyrogenation, of mostly salt solutions for hemodialysis, saccharide solutions and water (distilled, tap, subterranean),[46,49,50,56,57] the separation by sieving may have been aided considerably by adsorption phenomena. It was, for example, found that the retention of LPS was much higher when hydrophobic, nonfunctionalized polymer filters were used than with hydrophilic ones.[49,56,58] Moreover, polymer membranes exhibiting electropositive surface potentials have consistently shown higher adsorption capacity for LPS than others. It can be observed, for example, that unmodified Nylon (polyamide, PA) membranes, which tend to bear a slight positive surface charge, have a two orders of magnitude higher time-average LPS-removal factor than membranes of the same average pore size which bear no or negative surface charges. In fact, slightly negatively charged hydrophilic polymer membranes such as nitrocellulose (NC) or poly(vinyl acetate) (PVA) membranes do not adsorb LPS molecules at all or have very a small adsorption capacity, presumably due to a repulsion of the LPS molecule, which also bears a negative charge. However, at very low flow rates (extremely long contact times), even NC (pore size 0.22 mm) and PVA (0.2 mm) membranes did retain LPS, albeit with very low time-average removal capacities and time-average removal factors (0.09-0.14), presumably due to van der Waals-type interactions.[32,59]

The adsorption capacity of PVA membranes decreased with increasing average pore size, another indication that sieving is more important than adsorption for LPS removal in these membranes. A similar behavior was observed in the case of highly hydrophobic hollow-fiber polyethylene (PE) membranes (average pore size 0.15 x 0.6 mm and 0.04 mm). There too, the increase in the time-average removal factor from 0.13 to values between 76.5 and 90.0 observed for a decrease in pore size indicated that it is indeed the sieving mechanism, which is operative in LPS removal from water (tap, subterranean) with these membranes.[40,46,56,57] The highest values of the time-average removal factor observed for processes where sieving is clearly the operative principle for LPS removal are found with UF-type membranes. It is not surprising that in this case, the tendency is to observe an increase in the time-average removal factor with decreasing average pore size of the membrane. This fact indicates that if membranes do not contain functional charged groups, the size of the pores influences the LPS retention much more than the chemical nature of the polymer matrix. The time-average removal factor increased, for example, from something between 11.5 and 30 to something several orders of magnitude higher (between 31 940 and 13 690), when PEVA membranes with a molecular mass cut-off of 600 000, PAN, PS, PMMA and PEPA membranes with a molecular mass cut off of 20 000-30 000, and PAN and PS membranes with a molecular mass cut off of 6000 were used.[39,40]

It was shown, however, that approximately 55% and 47% of the LPS was found in the retentate after "depyrogenation" of distilled water with PAN and PS ultrafilters (molecular mass cut off of 6000), respectively.[39] The influence of the polymer nature on LPS retention is clearly seen in the depyrogenation of dialysate with hollow-fiber PAN, PS, PMMA and PEPA membranes (molecular mass cut off between 20,000 and 30,000). The PEPA membrane had the highest LPS retention efficiency for both *E. coli* 015 and *Actinetobacter calcoaceticus* produced LPS. The values of the time-average removal factor changed with the polymer nature in the following order: PEPA (32) > PMMA (30) ≈ PS (30) > PAN (26). In spite of the apparently similar values of the time-average removal factor for PMMA and PS membranes, the former were more effective in the LPS removal because often a lower absolute LPS concentration can be achieved with these membranes in the permeate. In this regard, only the PEPA membranes show a similar removal efficiency for LPS removal (*E. coli* 015 and *Actinetobacter calcoaceticus*) as the PMMA membranes. The adsorption capacity of PAN, PMMA and PS membranes may also depend on the LPS type. In the above-mentioned experiments, the concentration of LPS from *Actinetobacter calcoaceticus* in the purified dialysate was consistently 2-3 times higher than that of LPS from *E. coli* 015.

At present, the use of membrane filters with pore diameters smaller than the expected size of the contaminants are still considered "state-of-art" in depyrogenation. It was reported that

Table 5.2 LPS filtration with nonfunctionalized polymer membranes

Polymer	MW Cut-Off or Pore Diameter, μm	LPS-Containing Solution	LPS Concentration X10^{9}mg/ml		Time-Average		Ref.
			Initial	Final	Adsorption Capacity mg m^{-2}h^{-1}	Removal Factor X10^6 m h^{-1}	
		LPS Adsorption from Water					
NC	0.22	distilled water, 0.1 L	12,000	~12,000	0	0	66
PA	0.22	distilled water, 0.1 L	12,000	4800	55.5	19	
PE	0.15¥0.6	tap water, 2000 L	4710	90	0.012	0.13	57
PE	0.04	tap water, 13 L	3570	<10	0.059	76.5	56
		subterranean water, 13 L	420	<1	0.088	90	
PAN	6000	distilled water, 0.5 L	229,840	<0.16	5.0	31,940	39
PS	6000	distilled water, 0.5 L	260,210	<0.16	21.5	136,900	

continued on the next page

Table 5.2. Cont.

		LPS Adsorption from Buffer and Salt Solutions					
PA: PVA	0.45	20 mM PBS, pH 7	600,000	596,500	0.3	0.09	32
PVA	0.2	20 mM PBS , pH 7, 0.02 L	600,000	472,600	10.1	0.1	
PEVA	600,000	20 mM PBS , pH 7	600,000	2500	28.7	11.5	
NC	0.22	0.9% NaCl, 1 L	10,000	9200	0.105	0.14	59
PA	0.22	0.9% NaCl, 0.5 L	14,000	8000	4.6	13.5	66
PAN	20,000 – 30,000	dialysate, 1 L	8000	0.64 1.65	0.026 [a] 0.026 [b]	26 26	40

continued on the next page

Table 5.2. Cont.

PS	20 000 ~ 30 000	dialysate, 1 L	8000	1.03	0.03 [a]	30
				2.07	0.03 [h]	30
PMMA	20,000 ~ 30,000	dialysate, 1 L	8000	0.06	0.03[a]	30
				0.37	0.03[b]	30
PEPA	20,000 ~ 30,000	dialysate, 1 L	8000	0.06	0.032[a]	32
				0.06	0.032[b]	32

[66] Flat sheet filters: nylon (AMF/CUNO, Meriden, Conn., USA); nitrocellulose GS (Millipore Corp., Bedford, Mass., USA), 0.00039 m^2; LPS from *E. coli* O111:B4 in distolled water (0.1 L) and 0.9% NaCl (0.5 L); flow rate 50-80 mL/min. [57] Hollow-fiber chopped (1 mm) membranes PE-CF (Mitsubishi Rayon Co., Tokyo, Japan), 7.2 m^2; LPS in tap water (2000 L); flow rate 300 mL/min. [56] Microporous hollow-fiber EHF membranes (Mitsubishi Rayon Co., Tokyo, Japan), 0.007 m^2; LPS in tap water and subterranean water (13 L); flow rate 25 mL/min. [39] Hollow fiber membranes Microza PAN (1), Microza PS (2) (pall Corp., Clen Cove, NY, USA), 0.017 m^2; LPS from *E. coli* O113:H10 in distilled water; volume 0.5 L; pressure 80-90 kPa; premeate flow rate, mL/min: 14 (1), 45 (2). [32] Hollow fiber membrane PEVA (EVAL 4A (Kuraray Co., Osaka, Japan), 0.0025 m^2; hollow fiber membrane PVA (Kuraray Co., Osaka, Japan), 0.00151 m^2; flat sheet membrane Ultipor N66:PVA (Pall, Dreieich, Germany), 0.00134 m^2; LPS from *E. coli* 10498 (0.02 L) in 20 mM PBS, solutions differing in pH; concentration of bovine serum albumin (BSA) 1 mg/mL; flow rate 2 mL/min. [59] Flat sheet nitrocellulose GS membrane (Millipore Corp., Bedford, Mass., USA). 0.0017 m^2; LPS from *E. coli* O111:B4 in pyrogen-free 0.9% NaCl (1 L); pressure 10 kPa; flow rate 3.8 mL/min. [40] Hollow fiber membranes: FLX-15GW (PEPA), 1.5 m^2 (Nikkiso Co., Tokyo, Japan); BK-1.6U (PMMA), 1.6 m^2 (Toray Co., Tokyo, Japan); PNF-1700 (PAN), 1.84 m^2 (Asahi Co., Tokyo, Japan); PS-1.5UW (PS), 1.6 m^2 (Kawasumi Co., Tokyo, Japan); LPS from: a *E. coli* O15, b *Actinetobacter calcoaceticus* dissolved in dialysate (1 L); flow rate 100 mL/min.

membrane filters and hollow-fiber membranes with a molecular mass cut off between 6000 and 10,000 showed excellent results in depyrogenation of a number of liquids such as water and aqueous solutions of salts, sugars and some proteins.[40,60-65] Unfortunately, in practice this approach requires considerable financial investments because high-pressure pumps are required. In addition, these procedures are associated with a relatively high loss of product and tedious cleaning and rinsing procedures, especially when UF hollow-fiber membranes are used. UF-membrane type dead end filters have the advantage that they are generally disposable, and thus need not be cleaned and rinsed, their size easily can be easily adapted to the volume to be filtered, thus minimizing product loss, and that the apparatus required for membrane filtration is relatively simple and inexpensive. However, the small pore size of UF filters creates problems in this case because in many applications the pores tend to clog rapidly, resulting in low flow rates and throughputs. From an economical point of view, using this approach is limited to the depyrogenation of fairly small volumes of fluids or the removal of trace quantities from water and aqueous solutions on the laboratory scale. For the depyrogenation of large volumes, at high flow rates (gallons-per-minute), with a reduction to less than 10 LPS pg/mL, an approach based on adsorption rather than sieving seems to be more promising. Clearly, the development of filters retaining LPS by a combination of selective adsorption and filtration would offer a major advantage over other forms of depyrogenation media. Such filters could widen the particle size range and the type of material that could be removed by a given porous filter.

Membranes Based on Positively Charged Polymers for Endotoxin Removal

Historically, the first positively charged filter/adsorbers were depth filters based on asbestos fibers. In treatment of solutions, depth filtration with asbestos-containing filters has long been recognized as an effective method of pyrogen control.[33,36,66,67] Asbestos, an aluminosilicate (chrysolite), has a strong electropositive surface charge at a pH below 8.3. Thus the material is well fitted to adsorbing LPS molecules, which are negatively charged at a pH above 2. The usefulness of asbestos fibers in processing parenteralia has been ascribed to both its mechanical and its electrochemical properties.[66] However, the FDA has recently banned asbestos for use in the processing of small- and large-volume parenteralia because of a possible link between asbestos use and certain types of cancer.[68] The adsorption properties of many different (positively charged) adsorbents including kaolin, alumina, charcoal, kieselguhr, QAE Sepharose and some anion exchange resins have been checked in relation to depyrogenation after the FDA decision.[69-71]

LPS removal is extremely dependent on the environmental conditions. Studies of the depyrogenation of various liquids with depth filters (either bead or fiber based) have shown that the sorption of LPS using alumina, kaolin, perlite and diatomaceous earth depends strongly on the chemical nature and the composition of the liquid in question. Various additives may affect the size of the LPS molecule in water, i.e., through aiding or impeding aggregation. Inorganic cations (Ca^{2+}, Mg^{2+}) associate with the LPS molecules and aid the formation of high-molecular-mass aggregates, which are more easily removed by depth filtration. EDTA, on the other hand, will cause dispersion of LPS aggregates into smaller (even monomeric) units, thus making their removal with depth filter less effective. The strength of the interaction of the LPS molecule and the charged group of the adsorber surface is undoubtedly also influenced by the degree of ionization, both of the adsorber functional groups and those of the endotoxin molecule.

Cations in solution have a competing (inhibiting) effect on the adsorption of the negatively charged LPS molecules on positively charged surfaces. The inhibition increases with increasing valence of the cation. Depyrogenation with positively charged depth filters, made out of diatomaceous earth embedded in cellulose fibers, was found to also depend on both the pH and the NaCl concentration of the solution.[36,59,66] The small salt counter-ions compete successfully with the LPS molecules for active sites on the filter matrix; that leads to reduction in the LPS adsorption capacity or even, to a certain degree, displacement of the adsorbed LPS molecules from the matrix. For a number of membranes, the LPS adsorption capacity appears to peak around a pH of 6, which is one pH unit lower than the pK_a value of the functional groups of, for example, the Zeta Plus ZA cartridge.

However, in spite of difficulties with scale-up (increasing the filter diameter and the number of cartridge cells), it is now possible to process thousands of gallons of water at a high flow rate with depth filtration and to thereby reduce the LPS level to less than 50 pg/mL. Especially in large-scale applications, activated charcoal or charge-modified depth filters are still a popular means for depyrogenating of water, buffers (salt) and certain biological solutions required in drug preparation.[36]

The time-average removal factor of LPS from distilled water and salt solutions with depth filters based on diatomaceous earth is significantly higher (Table 5.3) than that with the previously mentioned nonfunctionalized polymer membranes (Table 5.2). The same is true for asbestos filters. However, when the pH of the water was increased from 4.0 to 8.5, the time-average removal factor decreased by a factor of 8. Changing the salt and glucose concentrations in the range of 0.9 to 5% and 5 to 50%, respectively, on the other hand, did not influence the efficiency of the LPS removal. The presence of NaCl (5%) in the glucose or serum solutions causes the time-average removal factor for the depyrogenation to deteriorate rapidly. This phenomenon is attributed to a possible competition between the LPS and the protein molecules for the surface binding sites. The time-average removal factors were also observed to depend on the LPS concentration itself. This may be connected to the possible adsorption of LPS aggregates, whose concentration increases with the LPS concentration.[72-74]

In spite of all their advantages, depth filters have the principal disadvantage that they tend to significantly contribute to the particle contamination of the fluid. Moreover, the loss of product may be unacceptably high when dealing with small-scale production (e.g., radiopharmaceuticals).[34] In some cases, an alternative may be found in the modified PA-based filter membranes Posidyne (0.2 mm) and Zetapor (0.22 mm). In contrast to the standard, uncharged PA membranes, these modified varieties bear positive surface charges and generally capture LPS from solutions very effectively.[41,66] They show a high time-average adsorption capacity and time-average removal factor, especially in depyrogenation of water and saccharide solutions, but their efficiency decreases in the presence of salts presumably due to an interference of the salt ions with the predominantly electrostatic interaction mechanism operative in LPS retention.

Membranes Based on Pseudoaffinity Ligands for Endotoxin Removal

During the last decades, functional proteins such as antibodies (IgG, obtained from human serum, ascites or hybridoma cell culture supernatants) have increasingly been considered for therapeutic applications as well as as reagents for in vivo diagnostics. Depyrogenation of these substances is extremely difficult and therefore also expensive. Not all membrane methods are suitable for the depyrogenation of LPS-contaminated proteins and IgGs. UF is not helpful since IgGs are relatively large proteins (MW 150,000) which would be retained along with the LPS aggregates. Adsorption on depth filters with activated charcoal or diatomaceous earth is usually accompanied by high loss in both the absolute amount of product and in specific activity.[34] Clearly the separation of LPS from these products requires a more sophisticated approach than a simple filtration through a depth filter.

Many investigations in biotechnology have shown that the utilization of an interaction between specific functional groups on both the adsorber and the target molecule improves considerably the selectivity of the retention of the target substance from complex, multicomponent biological fluids (principle of "affinity chromatography"). The affinity principle holds a

Table 5.3. LPS filtration with depth filters

Polymer	LPS-Containing Solution	LPS Concentration X10^9, mg/mL		Time-Average		Ref.
		Initial	Final	Adsorption Capacity, mg m^{-2}h^{-1}	Removal Factor X10^6,m h^{-1}	
LPS Adsorption from Water, Solutions of Salts and Saccharides						
PA 0.2μm (Posidyne)	distilled water, 0.93 L	32,400	a) 3:2	a)61.4	17,900	41
	0.9% NaCl, 0.66 L	45,500	b)15.8	b)78.6	5080	
	Ringer's solution, 0.2L	600	C)14.1	c)1.02	75	
PA 0.22mm (Zetapor)	distilled water, 0.5L	10,000	<50	76.8	1538	40
	5% glucose, 0.5L	-:-	<50	76.8	1538	
	0.9% NaCl, 0.5L	5000	500	34.8	77	
Diatomaceous earth	city water, 22710 L	500	50	0.11	880	36
	20 mM PBS, pH 7, 5 L	2000	19	0.5	27	
	20 mM PBS, 5 L	20,000	30	5.1	169	
	20 mM PBS, 5 L	200,000	42	50.7	1210	
Diatomaceous earth	tap water, pH4, 0.05 L	100,000	<30	769	25,640	40
	distilled water, pH 8.5, 0.05 L	-:-	240	767	3200	
	distilled water, pH 7.1, 10 L	200	<50	1.15	31	
	NaCl (0.9-5%), 0.05 L	100,000	<30	769	25,640	
	glucose (5%, 50%) 0.05 L	-:-	<30	769	25,640	
	5% glucose, NaCl, 0.05 L	-:-	40	769	19,200	
	5% serum, NaCl, 0.05 L	10,000	60	76.5	1280	
	5% glucose, 2 L	15,000	<50	115	2300	
Asbestos	distilled water, 0.1 L	12,000	156	91.1	592	

[41] Posidyne NAZ flat sheet nylon membrane, (Pall Europe, Portsmouth, U.K.); functional groups are not indicated; 0.0017 m^2; LPS from *E. coli* O55:B5 in pyrogen-free liquids: a) distilled water (0.93 L) b) 0.9% NaCl (0.66 L); c) Ringer's solution (0.03% KCl, 0.033% $CaCl_2$, 0.9% NaCl) (0.2 L); filtration under vacuum; flow rate 50 mL/min. [36] Commercial depth filter cartridge ZETA Plus ZA (CUNO Inc., Meriden, Conn., USA) composed from 9 depth filter cells; functional groups are not indicated; 0.839 m^2; laboratory module 0.0026 m^2; LPS from *E. coli* O111:B4; flow rate 3800 mL/min in city water (22710 L) filtration through commercial filter, 11 mL/min in buffer filtration through laboratory filter (5 L). [40] Depth filter 1D (AMF/CUNO, Meriden, Conn., USA); 0.00039 m^2; LPS from *E. coli* O111:B4; flow rate 50-80 mL/min. EKS asbestos filter (Seitz, Milldate, Conn., USA) 0.00039 m^2; LPS from *E. coli* O111:B4 dissolved in pyrogen-free distilled water and salt, glucose, glucose solutions; flow rate 50-80 mL/min.

unique place in bioseparation, since it constitutes the only possibility to separate almost any biomolecule on the basis of its own biological function rather than as a result of global physical or chemical properties. The fact that most affinity interactions tend to require mild, "physiological" conditions together with the typical extreme binding constants makes affinity interactions prime candidates for the removal of trace quantities of harmful contaminants from precious biotechnological or pharmaceutical products. In fact, affinity adsorption ("filtration") is one of the few methods which putatively can efficiently reduce the pyrogen concentration of various solutions to very low levels and under fairly mild conditions.[32,34,37,38,75-77] The reported apparent dissociation constants for some affinity ligands are in the order of 10^{-9} mol/L, which certainly ensures tight binding of the LPS molecule to the filter.[78]

The adsorption characteristics of these affinity membrane filters are principally determined by the nature of the ligands immobilized in the membranes. Up to now no suitable biospecific ligand has been found for LPS removal. Most affinity filtrations hence use so-called pseudoaffinity ligands. Several substances have been reported to mediate a high affinity for LPS, including adenine and cytosine, glycine (Gly) and histidine (His), histamine (Him), 1,6-diaminohexane (DAH), deoxycholate (DOC), polymyxin B (PMB), poly(L-lysine) (PLL), poly(ethyleneimine) (PEI), dextran, (hydroxyethyl)cellulose (HEC) and a-amylase.[34] A variety of chemical reactions are used to immobilize these (pseudo)affinity ligands on the filter membrane for the preparation of highly selective membrane adsorbers for depyrogenation.

The first investigations on the applicability of the affinity principle to depyrogenation of various solutions were made using particle-based supports. Ion exchange chromatography on DEAE provided some fairly interesting results for γ-globulin (antibody) depyrogenation although the LPS concentration was reduced only to 35% of the initial level.[79] In addition, a rather elaborate method for the preparation of LPS-free murine IgG_1 has been described which uses a multi-step procedure based on the combination of mixed-mode ion exchange chromatography and a QAE Sepharose column.[80] The IgG_1 recovery was about 80% and the LPS-concentration in eluate was below 0.25 EU/mL.

Membrane-based pseudoaffinity filters constitute promising alternatives to the established particle-based adsorbers. Affinity membranes combine the potential for high flux (low back pressure) of a membrane and the various process modes of filtration (e.g., cross and tangential flow) with the high affinity of a "column." This should result in faster and cheaper processes, which in addition are less limited in scale. Affinity filters are often more efficient than commercially available affinity gels carrying the same ligands, which is explained by faster mass transfer kinetics in the membranes due to the absence of pore diffusion, which is the main transport resistance in particulate systems.[81] Membrane systems thus possess certain advantages over particle-based ones, especially with regard to the depyrogenation of large volumes.

In contrast to nonfunctionalized polymer membrane filters, for which the LPS retention generally increases with decreasing pore size (see above), the opposite trend is observed with (pseudo)affinity membrane filters (Table 5.4).[32,34] In depyrogenation of phosphate buffer solution (20 mM, pH 7) using membranes with 1,6-diaminohexane (DAH) as ligand, the time-average removal factor increased from 20 to approximately 27,000 with increasing nominal pore size in the order: PEVA (molecular mass cut-off: 600,000) < PVA (average pore size: 0.2 mm) < PA (average pore size: 0.45 mm). If glycine rather than DAH was immobilized on the same membrane, the specificity of the LPS adsorption was very low. A similar phenomenon was observed when depyrogenation of buffers was attempted with PA and PVA membranes modified with a high-molecular-mass ligand, namely PEI (molecular mass 50,000). The values for the time-average removal factor were two orders of magnitude higher for PEI-activated PA membranes compared to the values obtained with PEI-activated PVA membrane. These results indicate that the chemical nature of the ligand but also spatial factors play a very important role in the affinity binding of LPS.

A simple way to test for selectivity of the LPS adsorption in (pseudo)affinity membranes calls for backflushing the membrane with pyrogen-free water after LPS adsorption. This has yielded LPS concentrations between 100 and 10,000 pg/mL, which is negligible in comparison

Table 5.4a. LPS filtration with (pseudo)affinity membranes

Polymer MW Cut-Off or Pore Size, μm	Ligand	LPS-Containing Solution	LPS Concentration X10^9, mg/mL		Time-Average		Ref.
			Initial	Final	Adsorption Capacity mg m^{-2}h^{-1}	Removal Factor X10^6,m h^{-1}	
		LPS Adsorption from Buffer Solutions					
PA, 0.45	Gly	20 mM PBS, pH 7	600,000	337,900	23.5	0.07	32
PVA, 0.2	Gly			459,700	11.2	0.02	
PEVA, 600,000	Gly			2200	28.6	13	
PA, 0.45	DAH			2	53.8	26,860	
PVA, 0.2	DAH			200	47.6	238	
PEVA, 600,000	DAH			1400	28.6	20.4	
				2	53.8	26,860	
PA, 0.45	PEI	20 mM PBS, 0.15 M NaCl	600,000	5	53.8	10,750	
		20 mM PBS in 1 M NaCl		100,000	44.8	0.45	
		20 mM PBS, pH 4.7		8	53.8	6716	
		20 mM PBS, pH 9.4		150	53.7	358	
PVA, 0.2	PEI	20 mM PBS, pH 7	600,000	166	47.6	287	
		20 mM PBS, 0.15 M NaCl		453	47.6	105	
		20 mM PBS in 1 M NaCl		100,000	39.7	0.4	
		20 mM PBS, pH 4.7		213	47.6	224	
PA 0.45mm	PLL	20 mM PBS, pH 4.7	621,100	6	55.6	9270	34
PA	PMB			3	55.6	18,540	
PA~Bis	His			12,400	54.5	4	
PA~DAH	His			9	55.6	6180	
PA~Bis	Him			8000	54.9	7	
PA~DAH	Him			440	55.6	126	
PA	DOC			21	55.6	2648	
PA~α-Amylase	DOC			6500	55.0	9	
PA	DEAE			2	55.6	27,810	
Cellulose	His	distilled water, 4 L	100,000	12[a]	0.44	36.8	37
		50 mM NaCl, 4 L		20[b]	0.18	8.8	
		50 mM NaCl, 3 L		32[c]	0.44	13.8	
		50 mM NaCl, 2 L		55[d]	0.71	12.8	
		100 mM NaCl, 1 L		55[e]	0.44	8	
		200 mM NaCl, 1 L		204[f]	0.44	2.2	

Table 5.4b. LPS filtration with (pseudo)affinity membranes

Polymer MW Cut-Off or Pore Size, μm	Ligand	LPS-containing Solution	LPS Concentration X10[9], mg/mL		Time-Average		Ref.
			Initial	Final	Adsorption Capacity, mg $m^{-2}h^{-1}$	Removal Factor X10[6], m h^{-1}	
		LPS Adsorption from Model BSA Solution					
PA, 0.45	PEI	BSA in PBS, pH 4.7	600,000	15	53.7 (96)	3580	32
		BSA in PBS , pH 7.2		25	53.7 (100)	2150	
		BSA in PBS, 0.15M NaCl		4000	53.4 (80)	13	
PVA, 0.2	PEI	BSA in PBS, pH 4.7	600,000	300	47.6 (92)	159	32
		BSA in PBS, pH 7.2		357	47.6 (92)	133	
		BSA in PBS, 0.15M NaCl		1770	47.5 (86)	27	
PA, 0.45	PLL	BSA in PBS, pH 4.7	661,000	10	59.20	5920	34
	PMB			10	59.20	5920	
	PEI			10	59.20	5920	
	DOC			330	59.17	179	
	DEAE			1	59.20	59,194	
PA-DAH, 0.45	Him			320	59.17	180	
	His			220	59.18	269	
		LPS Adsorption from Commercial Solution of Bovine Serum Albumin					
PA, 0.45	PLL	BSA in PBS, pH 4.7	6500	8	0.584	72.7	34
	PMB			8	0.584	72.7	
	PEI			8	0.584	72.7	
	DOC			22	0.576	26.4	
	DEAE			17	0.576	34.2	
PA-DAH, 0.45	His			117	0.568	4.89	
		LPS Adsorption from Commercial Solution of Lysozyme					
PA, 0.45	PLL	Lys in PBS, pH 7	13,400	270	1.176	4.4	34
	PMB			328	1.168	3.6	
	PEI			15	1.200	79.9	
	DOC			287	1.176	4.1	
	DEAE			347	1.168	3.4	
PA-DAH, 0.45	His			265	1.176	4.4	

with the LPS content of the original feed (600,000 pg/mL). Hence, sieving (e.g., together with compaction) and nonspecific adsorption of the LPS molecules can be ruled out. Instead, these findings indicate that the LPS-adsorption is indeed based mainly on a specific interaction of the endotoxin molecules with the ligands on the affinity filter.

Table 5.4c. LPS affinity with (pseudo)affinity membranes

Polymer MW Cut-Off or Pore Size, mm	Ligand	LPS-Containing Solution	LPS Concentration $X10^9$, mg/mL		Time-Average		Ref.
			Intial	Final	Adsorption Capacity mg $m^{-2}h^{-1}$	Removal Factor $X10^6$, m h^{-1}	
		LPS Adsorption from Commercial Solution of Mouse IgG_1 and Human IgG					
PA, 0.45	PLL	IgG_1 in AcB, pH 5.5	6250	33	0.56	16.9	34
	PMB			27	0.56	20.6	
	PEI			5	0.56	112	
	DOC			5	0.56	112	
	DEAE			21	0.56	26.6	
PA-DAH, 0.45	His			91	0.0552	6.1	
PEVA, 600,000	His	IgG in Tris, pH 7.4	25,500	11,500	0.408 (89)*	0.006	38
		IgG in PBS, pH 6	3900	1700	0.064 (97)	0.007	
		IgG in MES, pH 5	7560	4270	0.096 (95)	0.002	
		IgG in AcB, pH 5	22,800	4400	0.536 (99)	0.015	
PEVA, 600,000	His	IgG in AcB, pH 5	2500	10	0.056 (99)	4.9	
		IgG in AcB , pH 5	85,600	23,860	1.808 (79)	0.01	

[34] Microporous flat sheet membrane Ultipor N66 (Pall, Dreieich, Germany), 0.00134 m^2; LPS from *E. coli* DSM 10498 in 20 mM posphate buffer, pH 4.7, 7, 9.4; concentration of monoclonal antibody MAX 16H5 (mouse IgG_1) 1 mg/mL; volume of solution 15 mL, flow rate 2 mL/min. [38] Hollow-fiber membrane PEVA (EVAL 4A) (Kuraray, Osaka, Japan), 0.0041 m^2; LPS from DSM *E. coli* 10498 in buffers (Tris, phosphate, MES, acetate); volume of solution 15 mL; flow rate 0.2 mL/min; concentration of human IgG and mouse anti-human CD4 monoclonal antibody (IgG_1) 1 mg/mL. [37] Experimental membrane cartridge packed with affinity membranes containing His as a ligand on chemically modified microporous cellulose membrane (filter paper), 0.0017 m^2 x40 sheets; LPS from *E. coli* O55:B5 in: a distilled water (4 L); b (4 L), c (3 L), d (2 L) 50 mM NaCl; e. 100 mM NaCl (1 L), f. 200 mM NaCl (1 L); flow rate: a, c 5 mL/min, b 2 mL/min, d 8 mL/min, e,f 5 mL/min.

A comparison of the time-average removal factors for individual ligands gives rise to the assumption that electrostatic interactions play a major role in the adsorption mechanism (Table 5.5). Ligands, which show a very good potential for LPS removal, tend to carry a net positive charge. With an α-amylase support (net negative charge at pH 7), on the other hand, almost no reduction of the LPS concentration was achieved. The zwitterionic ligand Bis-His also did not support the removal of LPS very well. The structurally closely related but positively charged ligand Bis-Him (Fig. 5.2) on the other hand, was several orders of magnitude more efficient in this regard. Bis-Propa, which carries a methyl group instead of the imidazole moiety (Fig. 5.2) is even better than Bis-Him. However, the most effective endotoxin filters are obtained if the cationic spacer DAH is used to immobilize ligands of the His and Him types. This spacer is even (slightly) more effective without further modification.

Table 5.5. Electrostatic interactions and adsorption

Ligand/Adsorbant	pK	pI
Endotoxin	$pK_1 = 1.3$	-
	$pK_2 = 8.2$	-
Poly(L-lysine)	10.8 (NH_2)	9.7 (monomer)
Polymyxin B	>9 (NH_2 of DAB)	-
	~10.5 (NH_2 of DAB)	
Poly(ethyleneimine)	>10.5 (NH)	-
	>9 (terminal NH_2)	
Histidine	6.0 (imidazole)	7.6
	>9 (NH_2)	
	>10.5 (NH of DAH)	
Histamine	6.0 (imidazole)	-
	>10.5 (NH of DAH)	
Deoxycholate	>10.5 (NH of DAH)	-
Diethylaminoethyl group	9 – 9.5	-
α-Amylase	-	5.2
Bovine serum albumin	-	4.7
Lysozyme	-	11.2
Mouse immunoglobulin IgG_1 (MAX16H5)	-	5.5

The hypothesis of a strong contribution of electrostatic interactions to a "specific" LPS retention is further supported by the findings that the time-average removal factors of all LPS filters decrease dramatically in the presence of 1 M NaCl. For example, the time-average removal factor decreased from 26,860 and 287 to 0.4 for PEI-activated PA and PVA membranes, respectively, under these circumstances. Most likely, the electrostatic interaction, as a fairly long-range attraction, is responsible for bringing the LPS molecule into close proximity to the ligands. Once in close contact, the complex formation is strengthened by a number of short-range interactions, such as charge transfer, nonpolar and polar interactions or even hydrogen bond formation, resulting in the strong multifunctional binding type, which is typical for "affinity interactions". The harsh elution conditions required for filter regeneration (20% ethanol in 0.1 M NaOH) support the assumption of a multiforce attachment of LPS to the ligands. However, it is clear and often decisive for process development that electrostatic forces play a governing role in the primary interaction mechanism. Generally, very little adsorption is observed in the absence of positive net charges in the spacer-ligand system.

Variation of the pH changes the charge distribution for both the affinity ligands and the LPS molecules. At acid pH, the density of the negative charges in the LPS molecules decreases slightly in comparison to neutral pH, while the density of the positive charges increases for all ligands considered in Table 5.5. In some cases, the sign of the net charge of an affinity filter may even change as a function of the pH. For example, α-amylase becomes positively charged below a pH of 4.7. As a consequence, LPS molecules are attracted and bound by the immobilized enzyme, which is not possible at neutral pH due to electrostatic repulsion. Similar effects are observed with Bis-His. At pH 4.7, imidazole is protonated thus forming a positively charged structure (Fig. 5.2).

Under alkaline conditions, the situation is opposite to that in acidic medium. The charge density of the LPS molecules increases with increasing pH, while that of the ligands decreases. This usually leads to a considerable deterioration of the LPS removal in comparison to a neutral pH. Only affinity filters activated with the polymeric ligands PLL and PEI still achieved LPS-concentrations of less than 100 pg/mL in the permeate under these conditions. The time-average removal factor peaked at pH 7 (26,860) and decreased both for acidic (6716 at pH 4.7) and basic pH (358 at pH 9.4) for PEI-activated PA membrane adsorber. The same tendency was found for His-activated pseudoaffinity cellulose membranes.[37] When other ligands were immobilized on PA membranes, the efficiency of removing LPS from a 20 mM phosphate buffer decreased in the order: DEAE (F_T 27 810) > PMB (F_T 18 540) > PLL (F_T 9270) > DAH-His (F_T 6180) > DOC (F_T 2648) > DAH-Him (F_T 126) > α-amylase (F_T 9) > Bis-Him (F_T 7) > Bis-His (F_T 4). Decreasing the charge density of the LPS molecule itself has less impact on the LPS clearance than decreasing the charge density of the ligands. With polymer ligands PLL and PEI, this effect is less pronounced. Generally speaking, PLL and PEI can be used over a wide pH range, while ligands like Bis-His and α-amylase should be applied at pH values below their pI and pK, respectively.

The dominant role of electrostatic forces in the adsorption of LPS has important consequences for the handling of multi-component systems, such as contaminated protein solutions, where multilateral interactions based on electrostatic attraction or repulsion must be carefully balanced. Since the selectivity of this type of interaction is not very pronounced, the competition of proteins carrying a net negative charge at neutral pH, such as BSA, with the LPS molecules for binding sites on the affinity filter must be taken into account. Such a competition results in protein loss, while in addition being closely related to a considerable decrease in LPS removal (capacity). However, the separation of a protein and LPS can almost always be fine-tuned via the pH. The protein's interaction with the positively charged affinity filters can be reduced by adjusting the pH of the solution to the pI of protein. For example, the interaction of BSA (pI 4.7) with the ligands should become weaker with decreasing pH due to increasing protonation of the protein.

The pI of most proteins will be considerably higher than the pK_1 of the phosphate groups of the LPS molecule (pK_1 1.3). Hence, a pH can usually be found where the interaction of the LPS with the affinity filter is still possible, while the protein is no longer retained. For example, decontamination of BSA solutions at a pH of 4.7 instead of 7.0 did yield significantly better results (removal factors) in that the protein was successfully depyrogenated and at the same time quantitatively recovered at pH 4.7. In contrast, at pH 7, a loss of 20% of the target protein accompanied the LPS removal, which itself was also considerably less efficient. The effect was observed for all the investigated ligands, except DOC. At low pH, affinity filters with ligands such as PLL, PMB, PEI as well as DEAE could be used to reduce the LPS concentrations in the permeate to less than 15 pg/mL. Concomitantly, at least 96% of the BSA was recovered. Immobilized on PA membranes, these ligands also exhibited the highest values for the time-average removal factors in depyrogenation of model bovine serum albumin solutions (F_R = 5920 for PLL-activated, PMB-activated and PEI-activated PA membranes and F_T = 59194 for DEAE-activated PA membranes). However, in comparison with the control depyrogenation of a phosphate buffer solution, these values were 5-9 times lower.

Proteins carrying a net positive charge at neutral pH, such as lysozyme, are repelled from the positively charged affinity filters and thus do not compete directly with the LPS molecules for the binding sites. These proteins can, on the other hand, interact with LPS molecules, thereby neutralizing their respective charges. Two possible fates await the lysozyme/LPS complexes. If the interaction of LPS with lysozyme is much weaker than that with the affinity ligand in question, the protein/LPS complex will dissociate and most of LPS molecules will adsorb to the affinity membrane after all. If the protein/LPS complex is of similar stability or even stronger than the LPS/affinity ligand complex, the lysozyme will act as a carrier and most of the LPS will pass through the membrane and into the permeate. While it is often possible to more or less completely remove the LPS from solutions of acid proteins, such as BSA and ovalbumin, the performance of similar depyrogenation procedures tends to be much poorer in the case of basic proteins. It was concluded that this poor performance is due to the masking of the LPS activity by the respective protein rather than, for example, to an inhibition of the LAL test and that the lysozyme did indeed act as LPS-carrier.

When some of the highly selective affinity ligands for LPS (PLL, PMB, PEI, DAH-His, DOC and DEAE) were employed for the preparation of LPS-free solutions of two commercially available endotoxin-free proteins (BSA, Sigma A-3059 and lysozyme, Fluka 62971), most of them showed a considerable decrease in the time-average removal factor compared to results obtained with the model protein solutions prepared by adding known amounts of LPS to a priori LPS-free protein solutions. Only with PEI-activated membranes was it possible to reduce the residual LPS concentration of the lysozyme solution to 20 pg/mL. The value of the time-average removal factors obtained for the depyrogenation of commercial BSA, lysozyme, and human-IgG_1 solution with a PEI-activated PA membrane were very similar (72.7, 79.9 and 112 respectively), but decreased considerably when other ligands were used instead. From this point of view, only PEI meets the requirements of the international pharmacopoeia in regard to applicability at low initial endotoxin concentrations (~100 UE/m L). Not surprisingly, PEI is currently the most widely used ligand for affinity-based depyrogenation.

Some caveats may also exist, on the other hand, for the preparation of endotoxin-free protein parenteralia with LPS filters based on ligands such as polymyxin B and PEI, even though they effectively bind LPS from blood and plasma. If these ligands are present in the final plasma or IgG preparations even in trace amounts due to leaching of the affinity matrix, they may pose serious health threats to the patient. Polymyxin B is neuro- and nephrotoxic and PEI shows low biocompatibility.[38] In fact, of all the above mentioned affinity ligands, histidine is often considered "the best" for efficient depyrogenation of parenteralia, because it is biocompatible, safe and cheap.[37,38,82] In addition, histidine is known for its high chemical and physical stability, making this ligand an attractive choice for depyrogenation in spite of its somewhat lower time-average LPS removal factor when compared to ligands such as PMB and PEI.

It has been shown that LPS was preferentially adsorbed on a His-activated agarose-support in the presence of albumin and tumor necrosis factor (TNF).[83] The major disadvantage of using His as an affinity ligand for LPS, is that the molecule is also known to further the adsorption of different types of IgG under most operating conditions. As a pseudobiospecific ligand, this amino acid shows in fact affinity to a variety of therapeutic proteins including besides IgG, for example, also factor VIII. Recent studies have also revealed that supports with immobilized pseudoaffinity ligands, particularly with His, suffer in general from a considerable decrease in efficiency in the presence of proteins.[84] Hence, they are not in general applicable to the preparation of LPS-free protein parenteralia unless extensive preventive measures are taken.

The analysis of the present results shows that for efficient endotoxin removal preference should be given to adsorptive membrane having positively charged functional groups and pseudoaffinity ligands, which are especially effective in binding LPS from different sources and solutions. It can be said that the pH is the most important variable during the optimization of the depyrogenation of a pharmaceutical protein. The procedure has to be chosen in such a way that the target protein may be allowed to interact with the adsorber but certainly not with the LPS molecule (to avoid masking). At the same time, care should be taken that the interaction of the LPS molecules and the adsorber is not affected by these conditions. In this context, the choice of the ligands becomes important. For example, removal of LPS traces from human IgG (pI 9) would not be possible using Bis-His ($pK_{imidazole}$ 6). For the decontamination of acidic proteins, five ligands are available at the moment, namely the polymers PLL, PMB, PEI as well as DEAE and DOC (pH 7). His and Him, even when immobilized with DAH as spacer, show significantly worse clearance factors than the other ligands. In spite of the overwhelming use of this material, it should be recognized that the choice of a hydrophilic base polymer network for affinity filter production via ligand immobilization is not restricted to Nylon. Membranes based on regenerated cellulose or polysulfone can be modified by the same or slightly altered chemical procedures.

Membranes for Removal of Bacteria and Viruses from Aqueous Solution

Due to their high efficiency, ultrafiltration membranes have gained a well-deserved acceptance for the separation of various microorganisms from water as well as from buffer, salt and saccharide solutions. However, here too the large-scale production of parenteralia requires increasing the performance of the decontamination process, for example, in regard to throughput and scale. Currently an approach similar to the one taken for LPS removal, i.e., a combination of sieving and adsorption, appears to be very promising in the context of removal of microorganisms from aqueous solution. This can be achieved by using positively charged membrane adsorbers having a considerably higher pore size than the dimensions of microorganisms. Negatively charged filters with similar pore diameters tend to fail. This indicates that the retention of bacteria and viruses in these cases results from the electrostatic interactions of their negatively charged surfaces with the positively charged functional groups of the "filters".[36,59,66,75]

For example, the depth filters Zeta Plus ZA (surface-modified diatomaceous earth), positively charged polyamide Posidyne filters (NBZ, 0.45 mm; NHZ, 0.8 mm; NNZ, 1.2 mm) and hydrophobic hollow-fiber polyethylene membrane EHF have been used with good efficiency for the removal of organisms such as poliovirus type 1, bacteriophage MS-2, influenza virus and *E. coli* B 15597 from water (Table 5.6).[41] In these experiments, all Zeta Plus filters (0.2 to 2.0 mm) retained significant numbers of bacteria. However, the number of bacteria found in the filtrate did increase with the pore size of the membrane, whereas the time-average removal factor decreased by two orders of magnitude. It was observed that the bacteria were retained by a dual mechanism of sieving and adsorption.[41] Virus removal by the same depth filters, on the other hand, was mainly due to adsorption effects, since the dimensions of the viruses are several times smaller than the average pore size of the filters.

The feasibility of removing "particles" by adsorption is corroborated by the fact that the electropositive Zeta Plus S filters are also highly effective in the removal of latex beads, which are many times smaller than the filter pores, from solution. Even the large-pore Zeta Plus 05S filter is capable of efficiently removing 0.1 mm latex beads in such a situation. Results on particle filtration through depth filters indicated also that for particles larger than 1 mm, physical phenomena, such as mechanical sieving, are the most important for removal. For intermediate particles (0.5-1 mm), both sieving and electrokinetic mechanisms are of influence.[41] Finally, electrokinetic mechanisms are the dominating factor for removing particles of less than 0.1 mm in size, which is the case for most viruses. If the charges of the filter media and the particles are opposite, electrostatic attraction will force the particles to deposit within the filter. If the charges are of the same sign, repulsion will occur and deposition will be hindered. The Zeta Plus filters used in the investigation were all positively charged and most of the particles were negatively charged; hence, deposition occurred readily. The charge on the filter surface may be altered with changes in the pH, but according to the electric double layer theory of colloid chemistry, also depends on the electrolyte concentration in the solution being filtered.

Any particle, for example also a virus, immersed in an aqueous solution develops a surface charge by adsorbing ions on its surface. To maintain electrical neutrality, a diffuse layer forms around the particle, which is enriched in mobile counter-ions. This layer extends to some

Table 5.6. Membranes for removal of bacteria and viruses from aqueous solution

Adsorber	MW Cut-Off or Pore Diameter μm	Microorganisms	Concentration X 10^4, cfu*/mL		Time-Average			Ref.
			Initial	Final	Adsorption Capacity X10^8 cfu $m^{-2}h^{-1}$	Removal Factor X10^6 m h^{-1}		
Adsorption of Microorganisms from Tap Water, 2 L								
Diatomaceous	2	poliovirus type 1	10	2.6	31,320	163		59
earth		bacteriophage MS-2	100	100	0	0		
ZETA 05S		*E. coli* B 15597	1	0.3	3090	141		
Diatomaceous	1	poliovirus type 1	10	0.1	6090	618		
earth ZETA 50S		bacteriophage MS-2	100	8	56,700	77		
		influenza virus	100	0.01	61,950	61,800		
		influenza virus, pH 7	1	0.14	525	44		
		E. coli B 15597	1	0.001	620	6200		
Diatomaceous	0.2	poliovirus type 1	10	0.001	21,600	21,200		
earth ZETA 90S		bacteriophage MS-2	100	0.001	21,240	21,200		
		influenza virus	1	0.0001	216	21,200		
		E. coli B 15597	1	0.0001	216	21,200		
Adsorption Bacteria from Water and Salt Solutions								
PE	0.04	*E. coli* O113	300	0.0001	50^V	514	12,900	56
		E. coli O111:B4	450	0.0001	50	770	19,300	
		Salmonella minnesota	200	0.0001	50	343	8600	
PA (NBZ)	0.45	*Pseudomonas dimunita*	4	0.0001	a190	28.2	2820	41
					b10	27.6	2820	
PA (NHZ)	0.8	*Serratia marcescens*			a85	28.1	2820	
					b1	33.6	2820	
		Bacillus subtilis			a268	28.2	2820	
					b114	28.3	2820	
PA (NNZ)	1.2	*Serratia marcescens*			a9	28.0	2820	
					b0.5	28.8	2820	
		Bacillus subtilis			a1	30.0	2820	
					b2	30.0	2820	

^{V} breakthrough volume, mL; cfu – cell forming unit [59] Charged-modified cellulose diatomaceous earth depth filters (AMF/CUNO Division, Meriden, Conn., USA): ZETA Plus 05S, 0.0017 m^2; functional groups are not indicated; filtration under pressure 10 kPa; flow rate, 1200 ml/min; ZETA Plus 50S, 0.0017 m^2; flow rate 175 mL/min; ZETA Plus 90S, 0.0017 m^2; flow rate 60 mL/min; [56] Hollow-fiber polyethylene EHF membranes (Mitsubishi Rayon Co., Nagoya, Japan), 0.007 m^2; flow rate 0.5 mL/min; breakthrough volume 50 mL. [41] Positively charged nylon filters Posidyne (Pall Europe, Portsmouth, U.K.): NBZ, NHZ, NNZ; functional groups are not indicated; 0.0017 m^2; filtration under vacuum; flow rate 2 mL/min; a) water, b) Ringer's solution.

distance into the solution. If the concentration of the counter-ions in the bulk solution increases, for example due to the addition of salts or by increasing the pH, the thickness of the counter-ion enriched layer around the particle decreases, because a smaller volume will now contain enough counter-ions to neutralize the surface charge. The same is true for the charged surface of a filter. The reduction in the thickness of the diffuse layer facilitates the approach of two surfaces, allowing the van der Waals forces to show an effect.[85]

Electropositive filters are effective in the removal of a variety of microorganisms over a wide range of pH and salt concentrations, depending on the correlation between the surface charge of the filter and the microorganisms, respectively. On the other hand, electronegative filters of similar porosity are generally effective only in a limited range of pH and salt concentrations. All filters tend to perform best at low pH, but only electropositive filters perform well above pH 6. The electropositive Zeta Plus S filters retain far higher amounts of viruses than the negatively charged filters at a pH between 4 and 9. However, the capacity of the filters is always reduced at the higher pH values. Varying the NaCl concentration between 10^{-5} M and 0.1 M had no apparent effect on the removal of bacteriophage MS-2 by the electropositive Zeta Plus M filter. Lowering the pH, however, did slightly enhance the adsorption from 99.7% (pH 8) to 99.995% (pH 5). At near-neutral pH (tap water), Poliovirus was also adsorbed to a greater extent on electropositive than on electronegative filters.

High retention efficiency for *E. coli* O113, *E.coli* O111:B4 and *Salmonella minnesota* R595 was also exhibited by a hydrophobic hollow-fiber polyethylene membrane with a nominal pore size of 0.04 mm. Aqueous solutions containing bacterial cells were virtually bacteria-free after filtration through this membrane.[56] However, the values of the time-average removal factor were twice lower in this case compared to the positively charged Zeta Plus 90S filters in spite of the five times larger pore size of the latter. In retention of microorganisms with hydrophobic filters, sieving prevails apparently over adsorption.

In contrast to hydrophobic PE membranes, electrostatic forces were clearly involved in the retention of bacteria by positively charged polyamide Posidyne filters with nominal pore sizes between 0.45 and 1.2 mm. In this case the filter's efficiency depended to a large extent on the suspending medium. A smaller number of microorganisms was, for example, retained when a Ringer's solution was passed through the filter compared to a filtration from water. Breakthrough of bacteria into the filtrate was observed after only 20 ml of Ringer's solution were passed through the filter, while 1 L of water could be filtered under similar circumstances, before the breakthrough became noticeable. The data presented in Table 5.6 also show that the composition of the solution has a profound effect on the bacteria retention. The number of bacteria that could be effectively retained from solution with positively charged polyamide Posidyne filters depended in addition on a number of other operational parameters such as the effective filtration area, the construction of the filter, the total volume of the filtered liquid, the flow rate and the physicochemical properties of the solution. The viscosity of the solution, its pH and zeta potential, as well as the type and concentration of ionogenic admixtures did also influence the adsorption activity of these filters.

Membranes for Removal of Leukocytes from Blood Products

Leukocyte-depleted blood products, particularly red blood cell concentrates, are clinically used to avoid negative side effects in recipients after transfusion. Possible leukocyte-associated post-transfusion complications include human leukocyte antigen alloimmunization, graft-versus-host disease, platelet refractoriness, and transmission of viruses.[86] Amongst the various existing techniques for the selective removal of leukocytes from blood, filtration has become a popular method, because of its convenience and low costs.[87] Leukocyte filters have been specially developed for the purpose; they generally consist of fibrous materials made of Nylon,[88] PAN,[89] cotton wool,[90] cellulose acetate,[43] or polyester.[44,45] Currently available filters

Table 5.7. Membranes for removal of leukocytes from blood products

Adsorber	MW Cut-Off or Pore Diameter µm	Leukocyte Media	Concentration X10^4, cfu/mL		Volume, mL	Time-Average		Ref.
			Initial	Final		Adsorption Capacity X10^8 cfu m^-2h^-1	Removal Factor X10^3 m h^-1	
PU	25	granulocyte	100	19	70	a11.9	23	42
		suspension			50	b36.6	15	
PU~PEI	25	granulocyte	100	3		a14.2	49	
		suspension		31.5		b50.2	158	
PET		RBC leukocytes	750	60	500	101.1	18	
		granulocytes	400	52		51.0	11	
		lymphocytes	275	5.5		39.5	73	
		monocytes	50	0.8		7.2	92	
PET~PEI		RBC leukocuytes	750	49	500	102.8	22	
		granulocytes	400	40		52.7	15	
		lymphocytes	275	4.4		39.7	92	
		monocytes	50	0.3		7.2	244	

Cfu – cell forming unit; RBC – concentrate of red blood cells containing granulocytes, lymphocytes and monocytes. [42] Microporous polyurethane filters unmodified and modified (adsorption technique) with PEI 600 (MW 40000-60000); 0.0982 m^2; flow rate: a 2.4 mL/min, b 12 mL/min; Commercial Optima Cellselect leukocyte filters composed from poly(ethylene terephthalate) (NPBI, Emmer-Compascuum, Netherlands) unmodified and modified with PEI; 0.0982 m^2; flow rate: 2.4 mL/min.

capture more than 99% of the leukocytes from blood, while the loss in red cells is small.[46] The mechanism of leukocyte depletion by these filters is still under discussion, but the results of many studies indicate that the removal of leukocytes is accomplished by selective adhesion of leukocytes to the filter materials.[47,91] Recently it was, for example, shown that adhesion rather than mechanical sieving governs the leukocyte filtration through porous polyurethane (PU) membranes.[42,92] Attempts have been made to enhance the adhesion of leukocytes to PU and poly(ethylene terephthalate) (PET) membrane filters by their modification with PEI-ligands, since the cell-adhesive properties of PEI modified surfaces are well known in cell biology.[42]

A comparison of the adsorption efficiency of specially prepared PU and commercial PET (Optima Cellselect leukocyte) membrane filters (NPBI, Emmer-Compascuum, Netherlands) differing in the nominal pore size (38, 27, 22 and 7 µm) with those modified with PEI has shown that the adhesion of granulocytes to the modified membranes is always higher than to the unmodified ones (Table 5.7). More than 95% of the granulocytes (average size: 5 to 8 µm) are captured by PEI-modified filters (average pore size: 27 mm), whereas the depletion of cell numbers by unmodified filters is only 80%. The values of the time-average adsorption capacity and the time-average removal factor are approximately twice as high for the PEI-modified membranes. It was also found that the adhesive properties of these adsorptive membranes depend strongly on the cell nature. Certain leukocyte populations (such as lymphocytes) show three times lower adhesion to the same membranes than granulocytes. Unfortunately, the effect of PEI on leukocyte adhesion was only significant as long as red cells, platelets and plasma were absent from the liquid. There is virtually no difference in leukocyte adhesion from blood using PET or PEI-activated PET membranes. However, in the case of monocytes the time-average removal factor is about three times higher with the PEI-activated PET membranes than with the unmodified filters, when the filtration of a red blood cell concentrate (diluted to a hematocrit of ~ 60 with 0.9 % sodium chloride solution) was investigated.

Conclusions

Despite some remaining problems, for example with basic proteins, the new membrane adsorbers exhibit potential for various disciplines. Besides application in biotechnology (treatment of process waters, buffers, raw materials as well as the final products) and the pharmaceutical industry (treatment of parenteralia, diagnostics or dialysis buffers), the application of, for example, LPS-binding membrane adsorbers to therapeutic purposes, such as hemoperfusion in acute sepsis, is conceivable. A clear objective for future developments in the area of membrane adsorbers design will create an even higher selectivity for the target molecules. This will require the design of suitable ligands, most likely small molecules bearing positively charged groups, which are incorporated into the membrane matrix. Higher adsorption capacity of membrane adsorbers will hopefully be achieved by improving their porous structure.

References

1. Freitag R, Tennikova T High performance membrane adsorber chromatography (HPMC) for protein isolation. In: Aboul-Enein HY, ed. Analytical and Preparative Separation Methods for Biomolecules. Marcel Dekker, 1999:255-300.
2. Champluvier B, Kula MR. Microfiltration membranes as pseudo-affinity adsorbents: Modification and comparison with gel beads. J Chromatogr 1991; 539:315-325.
3. Klein E, Yeager DH. High-flux semipermeable membrane containing immobilized affinity ligands. US Patent 5,766,908, 1998.
4. Roper DK, Lightfoot EN. Separation of biomolecules using adsorptive membranes. J Chromatogr 1995; 702:3-26.
5. Bueno SMA, Haupt K, Vijayalaksmi MA. Separation of immunoglobulin G from human serum by pseudoaffinity chromatography using immobilized L-histidine in hollow fiber membranes. J Chromatogr 1995; 667:57-67.
6. Adachi T, Mogi M, Harada M et al. Selective removal of immunoglobulin E from rat blood by membrane-immobilized antibody. J Chromatogr 1995; 668:327-332.
7. Bueno SMA, Legallais C, Haupt K et al. Experimental kinetic aspects of hollow fiber membrane-based pseudoaffinity filtration: Process for IgG separation from human plasma. J Membr Sci 1996; 117:45-56.
8. Reif OW, Freitag R. Comparison of membrane adsorber (MA) based purification schemes for the down-stream processing of recombinant h-AT III. Bioseparation 1994; 4:369-81.
9. Tennikova TB, Freitag R. An introduction to monolithic disks as stationary phases for high-performance biochromatography. J High Resolut Chromatogr 2000; 23(1):27-38.
10. Freitag R, Splitt H, Reif OW. Controlled mixed-mode interaction chromatography on membrane adsorbers. J Chromatogr 1996; 728:129-137.
11. Kubota N, Miura S, Saito K et al. Comparison of protein adsorption by anion-exchange interaction onto porous hollow-fiber membrane and gel-packed bed. J Membr Sci 1996; 117:135-142.
12. Reif OW, Freitag R. Characterization and application of strong ion-exchange membrane adsorbers as stationary phases in high-performance liquid chromatography of proteins. J Chromatogr 1993; 654:29-41.

13. Lutkemeyer D, Bretschneider M, Buntmeyer H et al. Membrane chromatography for rapid purification of recombinant antithrombin III and monoclonal antibodies from cell culture supernatant. J Chromatogr 1993; 639:57-66.
14. Brief KG, Kula MR. Fast protein chromatography on analytical and preparative scale using modified microporous membranes. Chem Eng Sci 1992; 47:141-149.
15. Wang H, Li T, Zou H et al. The purification of human urinary kallikrein with ion exchange rapid flow membrane chromatography. Biomed Chromatogr 1996; 10:139-143.
16. Sarfert FT, Etzel MR. Mass transfer limitations in protein separations using ion-exchange membranes. J Chromatogr 1997; 764:3-20.
17. Gerstner JA, Hamilton R, Cramer SM. Membrane chromatographic systems for high throughput protein separations. J Chromatogr 1992; 596:173-180.
18. Josic D, Zeilinger K, Lim YP et al. Preparative isolation of glycoproteins from plasma membranes of different rat organs. J Chromatogr 1989; 484:327-335.
19. Gebauer KH, Thommes J, Kula MR. Breakthrough performance of high-capacity membrane adsorbers in protein chromatography. Chem Eng Sci 1997; 52:405-419.
20. Josic J, Reusch J, Loster K et al. High-performance membrane chromatography of serum and plasma membrane proteins. J Chromatogr 1992; 590:59-76.
21. Klein E, Yeager D, Seshardi R et al. Affinity adsorption devices prepared from microporous poly(amide) hollow fibers and sheet membranes. J Membr Sci 1997; 129:31-46.
22. Langlotz P, Kroner KH. Surface-modified membranes as a matrix for protein purification. J Chromatogr 1992; 591:107-113.
23. Strancar A, Barut M, Podhornik A et al. Application of compact porous tubes for preparative isolation of clotting factor YIII from human plasma. J Chromatogr 1997; 760:117-123.
24. Splitt H, Mackenstedt I, Freitag R. Preparative membrane adsorber chromatography for the isolation of cow milk components. J Chromatogr 1996; 729:87-97.
25. Klein E, Eichholz E, Yeager D. Affinity membranes prepared from hydrophilic coatings on microporous polysulfone hollow fibers. J Membr Sci 1994;90: 69-80.
26. Arica MY, Denizli A, Salich B et al. Catalase adsorption onto Cibacron Blue F3GA and Fe(III)-derivatized poly(hydroxyethyl methacrylate) membranes and application to a continuous system. J Membr Sci 1997; 129:65-76.
27. Kochan JE, Wu YJ, Etzel MR. Purification of bovine immunoglobulin G via protein G affinity membranes. Ind Eng Chem Res 1996; 35:1150-1155.
28. Nash DC. Comparison of diffusion and diffusion-convection matrices for use in ion-exchange separations of proteins. J Chromatogr 1998; 807:185-207.
29. Millesime L. Fractionation of proteins with modified membranes. Bioseparation 1996; 6:135-145.
30. Kasper C, Reif OW, Freitag R. Evaluation of affinity filters for protein isolation. Bioseparation 1997; 6:373-382.
31. Santarelli X. Characterization and application of new macroporous membrane ion exchangers. J Chromatogr 1998; 27:706 13-22.
32. Petsch D, Deckwer WD, Anspach FB et al. Endotoxin removal with poly(ethyleneimine)-immobilized adsorbers: Sepharose 4B versus flat sheet and hollow fiber membranes. J Chromatogr 1998; 707:121-30.
33. Sawada Y, Fujii R, Iqami I et al. The adsorption of endotoxin molecules in a microporous polyethylene hollow fiber membranes. J Hyg 1986; 97:97-102.
34. Petsch D, Beeskow TC, Anspach FB et al. Membrane adsorbers for selective removal of bacterial endotoxin. J Chromatogr 1997; 693:79-91.
35. Petsch D, Deckwer WD, Anspach FB. Proteinase K degestion of protein improves detection of bacterial endotoxins by the Limulus Amebocyte lysate assay: Application for endotoxin removal from cationic proteins. Anal Biochem 1998; 259:42-47.
36. Hou KC, Zaniewski R. Depyrogenation by endotoxin removal with positively charged depth filter cartridge. J Parent Sci Technol 1990; 44:204-209.
37. Guo W, Shang Z, Yu Y et al. Removal of endotoxin from aqueous solutions by affinity membrane. Biomed Chromatogr 1997; 11:164-166.

38. Legallais FB, Anspach SM, Bueno A et al. Strategies for depyrogenation of contaminated immunoglobulin G solutions by histidine-immobilized hollow fiber membrane. J Chromatogr 1997; 691:33-41.
39. Evans-Strickfaden TT, Oshima KH, Highsmith AK et al. Endotoxin removal using 6,000 MWCO polyacrylonitrile (PAN) and polysulfone (PS) hollow fiber membranes PDA.J Pharm Sci Technol 1996; 50:154-157.
40. Yamamoto C, Kim ST. Endotoxin removal by ultrafiltration through high-flux, hollow fiber membranes. J Biomed Mater Res 1996; 32:467-471.
41. van Doorne H. Sorption of bacterial endotoxin and retention of bacteris by positively charged membrane filters. J Parent Sci Technol 1993; 47:192-198.
42. Bruil A, Oosterom HA, Steneker I et al. Poly(ethyleneimine) modified filters for the removal of leukocytes from blood. J Biomed Mater Res 1993; 27:1253-1268.
43. Steneker I, Biewenga J. Histochemical and immunohistochemical studies on the preparation of leukocyte-poor red cell concentrates by filtration: The filtration process on cellulose acetate filters. Vox Sang 1990; 58:192-198.
44. Steneker I, Biewenga J. Histochemical and immunohistochemical studies on the preparation of leukocyte-poor red cell concentrates by filtration: The filtration process using three different polyester filters. Transfusion 1991; 31:40-46.
45. Snyder EL, DePalma L, Napychank P. Use of polyester filters for the preparation of leukocyte-poor platelet concentrates. Vox Sang 1988; 54:21-23.
46. Riverbi R, Menini C. Clinical efficacy of five filters specific for leukocyte removal. Vox Sang 1990; 58:188-191.
47. Kickler TS, Bell W, Ness PM et al. Depletion of white cells from platelet concentrates with a new adsorption filter. Transfusion 1989; 29:411-414.
48. Shettigar UR. Design analysis for an ultrafilter adsorber artificial kidney. Med Biol Eng Comput 1978; 16:15-24.
49. Homma JY. Bacterial, Endotoxin, Chemical, Biological, and Clinical Aspects. Basel: Verlag Chemie, 1984.
50. USP XXIII; Official Monographs. Rockville: US Pharmacopoeial Convention Inc., 1995.
51. Rietschel ET, Kirikae T, Schade FU et al. The chemical structure of bacterial endotoxin in relation to bioactivity. Immunobiology 1993; 187:169-190.
52. Anspach FB, Petsch D. Purification of murine IgG_1 on group specific affinity sorbents Bioseparation 1996; 6:165- 184.
53. Hannecart E, Dekegel D, Dupuydt F. Macromolecular structure of lipopolysaccharides from gram-negative bacteria. Eur J Biochem 1973; 38:6-13.
54. Shands J, Chun PW. The dispersion of gram-negative lipopolysaccharides by deoxicholate. J Biol Chem 1980; 255:1221-1226.
55. Rudbach JA. Some requisites in the system leading to hybrid formation between bacterial endotoxins. J Infect Dis 1970; 122:139-145.
56. Sawada Y, Fujii R, Igami I et al. Removal of endotoxin from water by microfiltration through a microporous polyethylene hollow fiber membrane. Appl Environ Microbiol 1986; 51:813-820.
57. Sawada Y, Fujii R, Igami I et al. Endotoxin removal from water using microporous polyethylene chopped fibers as a new adsorbent. J Hyg Camb 1986; 97:103-114.
58. Vanhaecke E, de Muynck C, Remon JP et al. Endotoxin removal by end-line filters. J Clin Microbiol 1989; 27:2710-2712.
59. Hou K, Gerba CP, Goyal SM et al. Capture of latex beads, bacteria, endotoxin and viruses by charge modified filters. Appl Environ Microbiol 1980; 40:892-896.
60. Muller K. Steam-sterilizable ultrafiltration for pyrogen-free water in pharmaceutical production. Swiss Contam Control 1990; 3(4a):146-148.
61. Nelson LL. Removal of pyrogens from parenteral solutions by ultrafiltration. Pharm Technol 1978; 2:46-49.
62. Validation Guide for Pall Microza Polysulfone 6,000 Molecular Weight Cut-Off Ultrafiltration Modules. Easy Hills: Pall Corporation, 1994.
63. Schindler R, Dinarello CA. Ultrafiltration to remove endotoxins and other cytokine-inducing materials from tissue culture media and parenteral fluids. Biotechniques 1990; 8:408-413.

64. Abramson D, Butler LD, Chrai S. Depyrogenation of a parenteral solutions by ultrafiltration. J Parent Sci Technol 1981; 35:3-7.
65. Cradock JC, Guder LA, Francis DL et al. Reduction of pyrogen—Application of molecular filtration. J Pharm Pharmacol 1978; 30:198-199.
66. Gerba CP, Hou K. Endotoxin removal by charge-modified filters. Appl Environ Microbiol 1985; 50:1375-1377.
67. Kaden H. The use of asbestos filter beads in the production of sterile and pyrogen-free solutions. Pharmazie 1975; 30:752-753.
68. Rossitto J. A solution of the asbestos problem. Pharm Technol 1979; 2:39-55.
69. Brasquet C, Roussy J, Subrenat E et al. Adsorption of micropollutants onto fibrous activated carbon: Association of ultrafiltration and fibers. Water Sci Technol 1996; 34:215-222.
70. Maitra SK, Yoshikawa TT, Guze LB et al. Properties of binding of *Escherichia coli* endotoxin to various matrices. J Clin Microbiol 1981; 13:49-53.
71. Nolan JP, McDevitt JJ, Goldman GS. Endotoxin binding by charged and noncharged resins. Proc Soc Exp Biol Med 1075; 149:766-770.
72. Pearson FR. Pyrogens, Endotoxins. LAL Testing and Depyrogenation. New York: Marcel Dekker, 1985.
73. Weinbaum S, Ajl SJ eds.. Microbial Toxins. Vol.4. New York: Academic Press, 1971.
74. Rietschel ET, Brade H. Bacterial Endotoxins. Sci Am 1993; 267:26-33.
75. Klein E. Affinity Membranes: Their Chemistry and Performance in Adsorptive Separation Processes. New York: John Wiley & Sons Inc., 1990.
76. Minobe S, Sato T, Tosa T et al. Characteristics of immobilized histamine for pyrogen adsorption. J Chromatogr 1982; 248:401-408.
77. Boester HD. Endotoxin specific membranes. Patent Application WO97/33683.
78. Minobe S, Watanabe T, Sato T et al. Characteristics and application of adsorbents for pyrogen removal. Biotechnol Appl Biochem 1988; 10:143-153.
79. Hou KC, Zaniewski R. Endotoxin removal by anion exchange polymeric matrix. Biotechnol Appl Biochem 1990; 12:315-324.
80. Neidhardt EA, Luther MA, Recny MA. Rapid, two-step purification process for the preparation of pyrogen-free murine immunoglobulin G_1 monclonal antibodies. J Chromatogr 1992; 590:255-261.
81. Thömmes J, Kula MR. Membrane chromatography in integrative concept in the down stream processing of proteins. Biotechnol Prog 1995; 11:357-367.
82. Minobe S, Watanabe T, Sato T et al. Preparation of adsorbents for pyrogen adsorption. J Chromatogr 1982; 248:401-408.
83. Mandjiny S, Vijayalakshmi MA, Rivat C et al, eds. Biotechnology of Blood Proteins. Colloque INSERM. Vol. 227. Montrouge: John Libbey Eurotext, 1993:189.
84. Matsumae H, Minobe S, Kindan K et al. Specific removal of endotoxin from protein solutions by immobilized histidine. Appl Biochem 1990; 12:129-140.
85. Kessick MA, Wagner RA. Electrophoretic mobilities of virus adsorbing filter materials. Water Res 1978; 12:263-268.
86. Rebulla P, Bertolini F, Parravicini A et al. Leukocyte-poor blood components: A purer and safer transfusion product for recipients? Transfusion Med Rev Suppl 1990 4(Suppl 1):19-23.
87. Meryman HT, Hornblower M. The preparation of red cells depleted of leukocytes – Review and evaluation. Transfusion 1986; 26:101-106.
88. Goldmann SF, Heiss F, Scheinert I. Removal of buffy coat from stored ACD blood by dextran agglomeration and subsequent filtration. Vox Sang 1973; 25:470-473.
89. Kayashima K, Asanuma Y, Smith JW et al. Simple on-line filters for the therapeutic removal of lymphocytes from blood. Trans Am Soc Artif Internal Organs 1981; 27:559-562.
90. Sirchia G, Parravicini A, Rebulla P et al. Effectiveness of red blood cells filtered through cotton wool to prevent antileukocyte antibody production in multitransfused patients. Vox Sang 1982; 42:190-197.
91. Wenz B. Clinical and laboratory precautions that reduce the adverse reactions, alloimmunization, infectivity, and possbility immunomodulation associated with homologous transfusions. Transfusion Med Rev 1990; 4(Suppl 1):3-7.
92. Bruil A, van Aken WG, Beugeling T et al. Asymmetric membrane filters for the removal of leukocyte from blood. J Biomed Mater Res 1991; 25:1459-1480.

CHAPTER 6

Stimulus Responsive Surfaces:

Possible Implications for Biochromatography, Cell Detachment and Drug Delivery

Igor Yu Galaev and Bo Mattiasson

Functional polymers, which can react, adjust or modulate their physicochemical character, i.e., in most cases their water-solubility, in response to an external stimulus, are generally referred to as "smart materials" or, if they are water-swellable, as "intelligent gels" or "wetware". The physical basis of this "smart" behavior is a fast and reversible change of the polymer microstructure from a hydrophilic to a more hydrophobic one, triggered by small changes in the environment. These microscopic changes result at the macroscopic level, for example, in the formation of a precipitate or in a change in the wettability of a surface to which the smart polymer is grafted. The changes are usually reversible and the system returns to its initial state when the stimulus is removed. The most commonly used environmental stimuli are temperature and pH, since the two variables are relatively easy to change. Other possible stimuli include electricity, ions, solvents, light and pressure. For biological processes, smart polymer systems, which respond to the presence of specific small molecules (such as glucose), are also highly useful.

Materials based on (or modified with) smart polymers, for example polymer-grafted surfaces, may revolutionize certain areas of biotechnology and medicine since the ensuing ability for self-recovery, self-adjustment or control, self-diagnosis, stand-by capability for detecting nonlinear onset, but also the ability to be externally tuned could be exploited in various ways. Table 6.1 summarizes a selection of possible applications of such materials in the life sciences.

The area of research on new and improved smart polymers is very active at present due to the obvious attraction of creating materials capable of carrying out increasingly complex tasks and performing various functions. However, we are still far from completely understanding the basic relationship between the chemical structure and the observed behavior. Unfortunately, research on water-soluble polymers, whether intelligent or not, has long been a somewhat neglected area of polymer chemistry. This is currently changing and this chapter intends to give a glimpse, sometimes based on just a very few existing examples, on what such materials may one day accomplish. It seems justified to consider various applications in which the highly nonlinear response of the polymer to small changes in the external medium is of critical importance for the successful functioning of the system as putative examples for the potential use of smart polymer in biotechnology and medicine. As most applications of smart polymers involve such phenomena as biorecognition and biocatalysis, which occur predominantly in aqueous environments, we will restrict ourselves to sufficiently hydrophilic or water-compatible polymers. Applications of smart polymers in organic solvents or hydro-organic mixtures are beyond

Synthetic Polymers for Biotechnology and Medicine, edited by Ruth Freitag.

Table 6.1: Application of smart polymers in the life sciences

Area	Application	Selected References
Medicine	Controlled drug release	35, 68, 71
	Site-selected drug delivery	82
Downstream processing	Cell-detachment	50, 51
	Affinity precipitation	4, 83
	Partitioning in aqueous polymer two-phase systems	84
	Modulated chromatography	41, 48, 64
Size-selective separation	Controlled permeation of membranes—chemical valve	14, 29, 56
	Concentrating macromolecular solutions	85
Analysis	Immunoanalysis	86
	Biosensors	87
Biocatalysis	Immobilization of living cells	88
	Enhanced mass-transfer	89
	Reversivbly soluble biocatalysts	90
	Feedback control of biocatalysts	91
Energy transducer	Artificial muscle	92

the scope of the chapter. We also decided to restrict the discussion to macromolecules physically attached or chemically grafted to a variety of surfaces. Systems, which are based on the polymers per se, either in solution or in the form of cross-linked networks or smart hydrogels, on the other hand, are not taken into consideration.

Stimulus-Responsive Polymers

The majority of the currently investigated stimulus-responsive polymer systems can be divided into three groups, those that change their microstructure upon a change in the pH, those that change it upon a change in temperature and those that are capable of reversibly forming cross-linked networks.” In the case of pH-sensitive polymers, the driving force behind the transitions is usually a neutralization of charged groups in the polymer by the pH shift. In the case of temperature sensitive polymers, the transition at elevated temperature has been ascribed to a decrease in the efficiency of hydrogen bonding to the surrounding water molecules under these conditions. Thermosensitive polymers often show some sensitivity to the ionic strength of their environment as well (hydrophobic effect, “salting out”). A number of reactions can be critical for (reversible) network formation. Below the base phenomena underlying “smartness” in polymers are discussed in more detail.

pH-Sensitive Polymers

The first group of smart polymers to be discussed here consists of macromolecules, usually weak polyelectrolytes, which depend upon their net charge for water solubility. Decreasing the net charge of the macromolecule can create poor solvent conditions in this case and concomitantly will reduce the electrostatic repulsion between the polymer segments. Other—attractive—forces become operative between the segments and the result is the collapse of the polymer coil often followed by an aggregation of the collapsed molecules. In a typical pH-sensitive smart polymer, the charge of the molecule is the result of protonation/deprotonation events (generally of carboxyl or amino groups) and thus depends strongly on the pH. Often it is possible to obtain a virtually neutral polymer by surpassing a certain pH. The pH-induced collapse/aggregation of pH-sensitive macromolecules tends to be very sharp and usually requires a change in pH of not more than 0.2-0.3 units. The replacement of some of the pH-sensitive carboxyl-groups with a noncharged and hence less hydrophilic group has been shown to increase the hydrophobicity of the copolymer resulting in transitions at higher pH.[4]

For instance, copolymers of methylmethacrylate and methacrylic acid undergo sharp conformational transition and collapse upon the acidification of the solution to a pH around 5, while copolymers of methylmethacrylate with dimethylaminoethyl methacrylate are soluble at low pH but collapse and aggregate under slightly alkaline conditions. The same is observed in the case of modified (i.e., more hydrophobic) cellulose derivatives with pending carboxyl groups, such as, for example, hydroxypropyl methyl cellulose acetate succinate.[5] A pH-sensitive polymer with reversible transition in the physiological pH-range (pH 7.0–7.5) was synthesized by copolymerization of N-acryloyl sulfametazine with N,N-dimethylacrylamide.[6]

Thermosensitive Polymers

The second important group of smart polymers consists of thermosensitive molecules. These are typically uncharged polymers soluble in water due to the hydrogen bonding of groups in their segments with the water molecules. The efficiency of hydrogen bonding decreases in general with increasing temperature. Consequently, there is a critical temperature for some polymers at which the efficiency of hydrogen bonding becomes insufficient to maintain the macromolecule in solution (contact segment/segment preferred over contact segment/water); the macromolecule collapses and aggregation/precipitation usually take place. It can be said that thermosensitive polymers undergo phase transition because they become progressively more hydrophobic with the increase in temperature. At some point the hydrophobic interactions between polymer molecules become more favorable then polymer-water interactions and the polymer molecules collapse. Hydrophobic interactions in general are promoted by high salt concentrations ("salting out" effect), which in the case of thermosensitive polymers are known to shift the transition to lower temperatures.[7] Concomitantly, hydrophobic interactions with the solvent are promoted and the critical temperature is elevated (or altogether eliminated) by the addition of organic solvents, detergents and chaotropic agents. As with the pH-sensitive polymers, the composition of the aqueous solution plays a major role in determining the behavior of thermosensitive smart polymers.

Most of the thermosensitive polymers studied worldwide fall into two groups. These are:

1. Poly(N-alkyl substituted acryl amides) and amongst them the most important certainly, poly(N-isopropylacrylamide) (poly(NIPAAM)), with a transition temperature of 32° C[8] and
2. Poly(N-vinylalkylamides), for example, poly(N-vinylisobutyramide), with a transitiontemperature of 39° C[9] or poly(N-vinyl caprolactam) with a transition temperature of 32-33° C[10] (depending on polymer's molecular mass)

Modern polymer chemistry provides in addition a variety of other polymers with different transition temperatures ranging from 4-5° C for poly (N-vinyl piperidine) to above 100° C for poly(ethylene glycol).[11]

Other than the pH-sensitive polymers, where the carboxyl or amino groups responsible for the charge can also be used for covalently linking the polymer to a surface, thermosensitive polymers do not necessarily contain such inherent reactive groups. It is thus often necessary to specifically introduce such groups during the polymer synthesis, for example through copolymerization of suitable monomers. N-Acryloylhydroxysuccinimide or glycidyl methacrylate are popular comonomers for NIPAAM in this context since they allow subsequent coupling of the polymer to hydroxy- or amino-groups.[12,13] Copolymers containing methacrylic or acrylic acid units have also been coupled to hydroxy- or amino-groups on the surface via appropriate activation procedures.[14] The introduction via a copolymerization has the advantage of allowing some control over the coupling chemistry. This is even most pronounced in cases where only a single, usually terminal coupling site is introduced. Examples include some thermosensitive polymers prepared by group[15] or chain transfer[16] polymerization ofrother approaches where the active group is incorporated via the initiator of the polymerization.[17]

The critical temperature of a given thermosensitive (co)polymer can often be fine-tuned by adding more hydrophilic or more hydrophobic comonomers to the structure. The former increases the transition temperature of the polymer while hydrophobic comonomers have the opposite effect.[18] The inclusion of comonomers with pH-depending hydophobicity (i.e., charge-mediated) adds to flexibility of the system. For example, at pH 8.0 where imidazole moieties are noncharged and hence relatively hydrophobic, copolymers made of NIPAAM and vinyl imidazole undergo phase transition at about 35° C, i.e., at a temperature slightly above that observed for the poly(NIPAAM)-homopolymer (32° C). When the pH is lowered to 6, the imidazole groups become protonated and hence very hydrophilic. As a consequence, no transition occurs even when the polymer solution is heated to 80° C.[19]

The properties of smart polymers, which are important for biotechnological and medical applications, often can be controlled not only by the comonomer composition, but also by the polymer architecture. For example, block copolymers with a thermosensitive "smart" part consisting of poly(NIPAAM) do not separate from aqueous solutions in the same manner as random copolymers do. In particular, a reversible gel rather than a concentrated polymer phase is formed in response to an increase in temperature.[20] Comb-like copolymers with poly(NIPAAM)-grafts show a faster and more pronounced response to changes in temperature as compared to random copolymers.[21,22]

Reversibly Cross-Linked Polymer Networks

The third type of smart polymers is based on systems, where the (soluble) polymer molecules are reversibly (noncovalently) cross-linking into (insoluble) polymer networks. The networks either precipitate or form physical gels. The most common of these systems are based on polymers with sugar-ligands, which can be cross-linked by lectins with multi-binding sites (for example concanavalin A) and on polymers using the well-known boronate-polyol interaction to achieve this goal.[23-26] The reversibility of the response is achieved through the addition/removal of low-molecular weight analog of the polymer. For example, when small sugars with high boronate affinity are added to the boronate-crosslinked polymer, they will compete with the polymer for the boronate groups, thereby destroying the intra-polymer cross-links and thus disengaging the network.

Polymer-Grafted Surfaces

Smart polymers, both of the pH- and the thermosensitive type, have been covalently coupled or "grafted" to solid surfaces thereby endowing them with new properties. Surface modification through physical adsorption is also possible. Covalent grafting to the solid support is usually achieved by letting polymers with terminal (only single-point attachment possible) or randomly distributed (multipoint attachment possible) active groups react with corresponding reactive groups on the surface in question. Another way, especially to achieve single point attachment of smart polymers, is to first covalently modify the surface with a polymerization initiator and then to carry out the polymerization of the monomer from these starting points. The growth of the polymer chains occurs only at the sites where the initiator was coupled.[27] Alternatively, the solid support can be irradiated by light or plasma beam in the presence of the monomer.[28,29] Irradiation results in the formation of active radicals on the surface which are capable of initiating the polymerization. The irradiation methods result in a high density of grafted polymer, but the polymerization is less controlled than in the case of covalently coupled polymers or polymers produced from covalently coupled initiators. Cross-linked polymer gels often form when irradiation is used, especially in combination with high monomer concentrations or in the presence of a cross-linker.[22,30]

For some applications it may be desirable to anchor a smart polymer to a lipid membrane. In this case, an appropriate "anchor" is required to securely link the polymer to the membrane. Such lipophilic anchors have been introduced into macromolecules, for example, by copolymerization of poly(NIPAAM) with comonomers having large hydrophobic tails like N,N-didodecylacrylamide,[31] by using a lipophilic radical initiator,[32] through the modification of the copolymer following polymerization[33] or by modifying polymers carrying a suitable active endgroup with a phospholipid.[34] Alternatively, it has been suggested to couple the polymer covalently to an active group found within the molecular structure of the lipid, which formed the membrane in question.[35]

When grafted to a surface, the polymer molecules in question can no longer aggregate; however, the conformational transition from a more hydrophilic to a more hydrophobic state (collapse of the polymer coil) still takes place and thus allows to regulate the hydrophobicity of the surface. A surface bearing grafts of smart polymers will be hydrophilic when the polymer molecules are in the expanded "soluble" form and hydrophobic when the polymer is in the collapsed "insoluble" form. The change in hydrophobicity of a surface grafted with poly(NIPAAM) molecules was demonstrated by contact angle and water absorbency measurements.[36-39] The transition temperature of adsorbed (presumably via multipoint attachment) poly(NIPAAM) molecules has been observed to be lower than in bulk solution. In addition, the properties of the layer of the collapsed macromolecules formed above the transition temperature seem to depend strongly on speed of the temperature increase. When the temperature is increased slowly, a more "liquid-like" polymer layer is formed while the layers formed at high speeds have more "solid-like" properties.[40]

Temperature-Responsive Chromatography

The idea to change the structure and adsorptivity of polymer-grafted surfaces by a slight change of the temperature or the pH is very attractive, because a drastic change in the interaction (for example adsorptivity) of this surface with a given solute may thus be achieved. One area, which depends strongly on the controlled interaction of solutes with adsorptive surfaces is chromatography. In this context first encouraging results with "smart" columns have been published. For example, the chromatographic separation of some steroids and drugs, using HPLC-columns packed with poly(NIPAAM)-grafted beads, was strongly dependent on the temperature with a steep increase in both retention and resolution when the temperature was increased from 5° C to 50° C.[41,42] Especially the retention times of the more hydrophobic steroids (hydrocortisone acetate, testosterone) were significantly longer at higher temperature. The retention times observed for the reference column packed with nonmodified (nongrafted) beads, on the other hand, were much shorter and decreased in fact with increasing temperature. At low temperatures retention was preferably through hydrogen bond acceptors, while hydrophobic interactions dominated the retention of the solutes at higher temperature. The temperature increase brought about a similar change in the retention behavior of the poly(NIPAAM)-grafted column as the more common change in the composition of the mobile phase (addition of organic solvents to aid solute elution) for the nongrafted reference column.[43]

The architecture of the graft macromolecules was observed to significantly affect the temperature response of the poly(NIPAAM)-modified stationary phase. The surface wettability of the poly(NIPAAM)-grafted surfaces changed dramatically with temperatures between 32° C and 35° C (corresponding to the phase-transition temperature for poly(NIPAAM) in aqueous media) in cases, where the poly(NIPAAM) was grafted via the end group either directly to the surface or to a looped chain copolymer of NIPAAM and N-acryloylhydroxysuccinimide, already there. The loop-grafted surface itself also showed a pronounced change in wettability, but at slightly lower temperatures (~27° C). This reduction in transition temperature is most likely due to a loss in the freedom of conformational transitions in case of the multipoint-attached macromolecules. Amongst the three surfaces the one with a combination of both loops and terminally grafted chains showed the largest surface free energy change upon phase transition.[44] To reduce the response temperature of the chromatographic matrix, a copolymer, namely poly(NIPAAM-co-butyl methacrylate) with a transition temperature of 20° C was grafted to the stationary phase instead of poly(NIPAAM)-homopolymer chains. A mixture of insulin chain A, β-endorphin fragment 1-27 and insulin chain B was successfully separated on this polymer-grafted matrix at 30° C where the copolymer was in a collapsed hydrophobic state. A separation of these polypeptides, consisting of 21–30 amino acid residues each, was not possible at 5° C, i.e., below the transition temperature when the copolymer was in the extended hydrophilic conformation.[45]

Less pronounced changes in the retention on poly(NIPAAM)-grafted stationary phases above and below the critical temperature were observed in the case of large protein molecules like immunoglobulins. About 20 % of the protein molecules adsorbed to the poly(NIPAAM)-grafted matrix at 37° C (i.e., above the critical temperature). These were eluted by decreasing the temperature to 24° C (i.e., below the critical temperature).[46] Although the protein adsorption to the poly(NIPAAM)-grafted surfaces could be controlled to some extent, a quantitative elution of adsorbed proteins by a temperature shift has not yet been demonstrated. On the other hand, a combination of temperature-responsive polymeric grafts with biorecognition elements, such as affinity ligands, proved to be successful strategy for temperature controlled protein chromatography. In this context, a second feature of surface attached stimulus-responsive polymers may become interesting for chromatography. In addition to a change in the hydrophobicity of the surface, these molecules may in their extended (hydrophilic) form, simply block the access of the analyte molecules to the interactive sites on the matrix surface. For example, poly(N-vinyl caprolactam), a thermoresponsive polymer with a critical temperature of about 35° C interacts efficiently with the triazine dye, Cibacron Blue, a widely used ligand for a dye-affinity chromatography of various nucleotide dependent enzymes.[47] The polymer is known to bind strongly via multipoint interaction to the dye ligands. At elevated temperature the polymer molecules are in a compact globular conformation and therefore will bind (and block) only a few ligands. Enzyme molecules, such as lactate dehydrogenase from porcine muscle, have good access to the majority of the ligands and thus interact strongly with the surface. When the temperature is decreased, the polymer molecules undergo transition to the more expanded coil conformation. The polymer molecules interact now with more ligands and begin to compete with the bound enzyme for the ligands. In the end, the enzyme may be completely displaced by the expanded polymer chains.[48] The principle of temperature-induced elution of proteins is illustrated in Figure 6.1. In fact, a system has been proposed, where only a change in temperature was sufficient to elute the protein from the column (Fig. 6.2). No

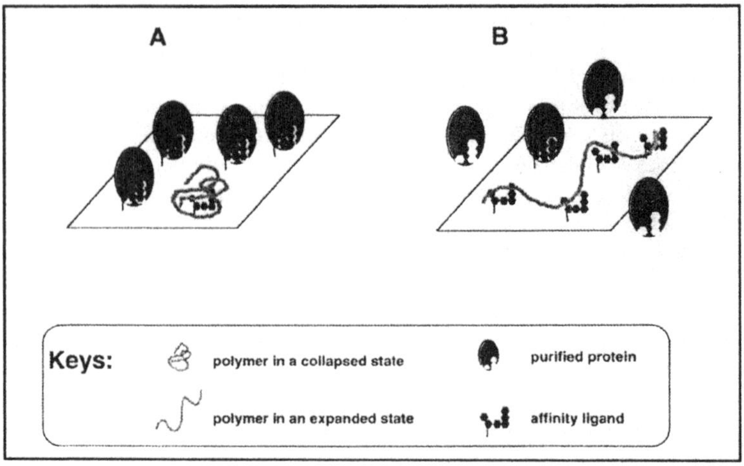

Figure 6.1. Schematic illustration of temperature-induced elution. A—at elevated temperature adsorbed polymer is in a collapsed state and interacts with one ligand making the other available for binding of purified protein; B—at low temperature adsorbed polymer is in an expanded state and interacts with a few neighboring ligands displacing bound proteins in solution.

additional changes in the buffer composition were necessary. Using temperature changes as eluting factor is quite promising because it eliminates the otherwise unavoidable step of separating the target protein from the elution mediator, in dye-affinity chromatography usually a competing nucleotide, or high salt concentrations.

Cell Detachment from Polymer-Grafted Surfaces

Cells adhere and grow more easily on hydrophobic surfaces than on hydrophilic ones. In fact, in accordance with thermodynamic expectations, the tendency for adhesion decreases with increasing surface tension of the support.[49] While a strong attachment of the cells is desired during growth, this creates problems once the cells need to be—gently—removed from the surface, for example for passaging or final applications. Traditionally, cells are detached form culture supports by protease treatment, which indiscriminately hydrolyzes various membrane-associated protein molecules together with the proteins responsible for surface attachment and thus often damages the cells. A technique for gentle cell detachment, which does not require the use of enzymes, is therefore of great interest to cell biologists. The idea that the changes in the hydrophobicity of a poly(NIPAAM)-grafted surface from strongly hydrophobic above the critical temperature to more hydrophilic below it may be used in this context is not too farfetched. In one successful experiment, poly(NIPAAM) was grafted to polystyrene-based culture dishes using the electron beam technique to activate the surface.[50] Bovine hepatocytes, i.e., cells known for their extreme sensitivity to enzymatic treatment, were cultivated on these dishes for 2 days at 37°C. The poly(NIPAAM)-grafted surface was expected to be hydrophobic at this temperature, since it is well above the critical temperature of the graft-polymer. Cell growth was not impeded by the graft, as the cell numbers on both the poly(NIPAAM)-grafted and the control dishes were nearly identical. The cells were subsequently detached by incubation at 4°C for 1 h. In case of the polymer-grafted dishes, this decrease in temperature rendered the surface much more hydrophilic and the cells were literally "pushed off" the surface by the expanding polymer molecules, Figure 6.3. Nearly 100% of the hepatocytes were detached and recovered from the poly(NIPAAM)-grafted dishes at low temperature, while only about 8% of

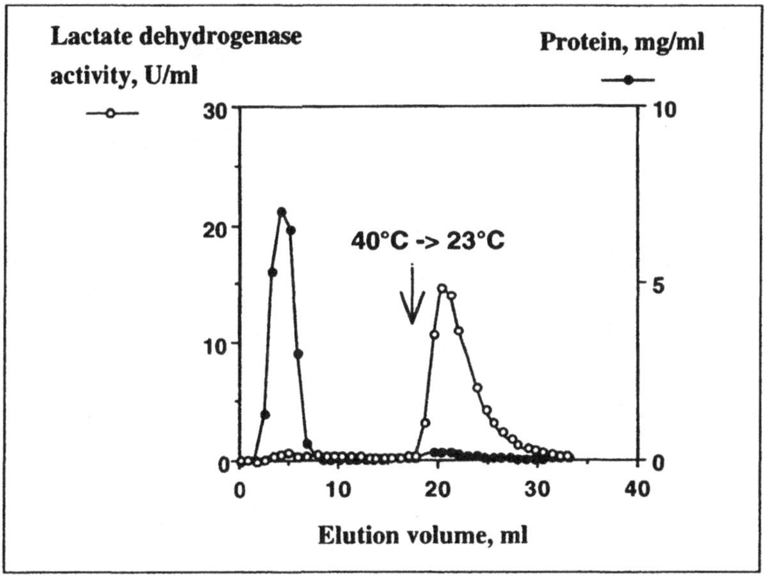

Figure 6.2. Elution profile of lactate dehydrogenase from porcine muscle and total protein with 0.1 M KCl from a Blue Sepharose column with adsorbed poly(N-vinyl caprolactam) (a thermosensitive polymer with transition temperature about 35°C). The crude extract was applied on a column at 40°C. Arrow indicates when the column was cooled to room temperature and elution was continued at this temperature. Experimental conditions: column 1.3 x 0.9 cm I.D., flow rate 0.17 ml/min, fractions collected every 5 min. Redrawn from[48].

the cells were detached from the control dish. The technique has since been extended to a variety of different cell types.[51]

It should be noted that the hepatocytes recovered at 4° C from the grafted surfaces retained their native form with numerous bulges and dips and did reattach well to a hydrophobic surface including the grafted polymer when the temperature was increased above the one required for the conformational transition of the poly(NIPAAM)-graft. Conversely, the enzyme-treated cells typically had a smooth outer surface and lost their ability to attach themselves to any surface to a considerable degree. In fact, only 14% of the hepatocytes recovered by protease (trypsin) treatment contrary to 73% of the ones recovered from poly(NIPAAM)-grafted dishes, were able to attach again to a culture dish. In addition, the hepatocytes were recovered from poly(NIPAAM)-grafted dishes without loosing their intercellular junctions, i.e., in an assembled state, which is known to be critical for their viability. All in all, cells recovered by a temperature shift from poly(NIPAAM)-grafted surfaces showed an intact structure and maintained normal cell functions.

Temperature responsive cell culture dishes have also proven useful for investigating the molecular machinery involved in cell-surface detachment.[52] At 37° C, grafted and nongrafted surfaces showed no difference with regard to the attachment, spreading, growth, confluency and morphology of bovine aortic endothelial cells. Stress fibers, peripheral bands and focal contacts were established in a similar manner. Differences between the two surfaces became apparent when the temperature was lowered to 20° C. In case of the poly(NIPAAM)-grafted surface, the previously spread-out cells lost their flat morphology and acquired a more rounded appearance similar to that of cells immediately after plating. A mild agitation of the dish was sufficient to make the cells float free from the poly(NIPAAM)-grafted surface without any

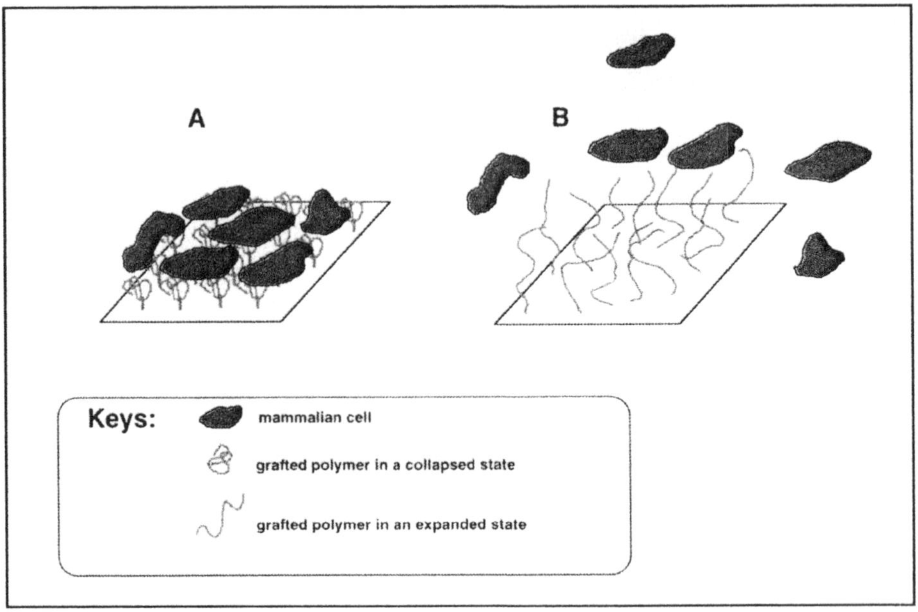

Figure 6.3. Schematic illustration of temperature-induced detachment of cells. A – at elevated temperature grafted polymer is in a collapsed state, the surface is hydrophobic and cells attach to it; B—at low temperature grafted polymer is in an expanded state, the surface is hydrophilic and cells detach easily from the surface.

trypsin treatment. Neither the cell morphology nor the cell detachment was found to change upon the temperature switch in the case of the nongrafted surfaces. The addition of an ATP synthesis inhibitor, namely sodium azide, or a tyrosine kinase inhibitor, namely genistein, suppressed the cell morphology changes and the cell detachment, while the addition of a protein synthesis inhibitor, namely cycloheximide, slightly enhanced the cells tendency for detachment. An actin filament stabilizer, namely phalloidin, and its depolymerizer, namely cytochalasin D, also inhibited cell detachment. These findings suggest that cell detachment on grafted surfaces (and perhaps to hydrophobic surfaces in general) is mediated by changes in intracellular signal transduction cascades and accompanied by a reorganization of the cytoskeleton.

Perhaps the most important application of the temperature-induced detachment of cells will be in the area of producing artificial organs, and more precisely, artificial skin. For such applications skin cells are grown on a culture dish from which they need to be detached keeping as much as possible of the intercellular structure intact before application to the patient. When using a thermosensitive growth-support, the cells are detached not as a suspension of more or less individual cells (the usual result of protease-induced detachment) but preserving their intercellular contacts to a large extent. For example, fibroblasts were cultured at 37° C on a poly(NIPAAM)-grafted collagen support until the cells completely covered the surface.[50] Afterwards the temperature was decreased to about 15° C. A few minutes later, the fibroblasts started to detach themselves from the dish as a sheet, and after about 15 min a completely detached cell sheet floated in the culture medium. In a similar experiment using a poly(NIPAAM)-grafted support, retinal pigmented epithelium cells were able to restore in vitro an environment similar to that found in vivo in that they formed even a tight junction barrier membrane once confluency was reached at 37° C.[53] This was supported by the concomitant changes in morphology, transepithelial resistance and directionally specific fluid transport. Once the temperature was lowered to 20° C, the cells detached as an oriented polarized sheet. The detached cells could subsequently be transplanted to another culture surface without functional and structural changes. In yet another approach, grafting of poly(NIPAAM) onto polystyrene surface by a photolitographic technique was used to create a special pattern on the surface and cultured mouse fibroblast STO cells were shown to grow on the entire surface, but to detach upon lowering of the temperature only from where the poly(NIPAAM) had been grafted.[54] Patterned ultrathin films of smart polymer may thus constitute interesting substrates for controlling the interactions of cells with surfaces and even be capable of directing the attachment and the spreading of cells.[55]

Controlling Porosity via Smart Polymers—The "Chemical Valve"

Stimulus-responsive polymer molecules should have a much larger size in the expanded hydrophilic state compared to the more compact hydrophobic one. When such a polymer is grafted to the surface of a pore, for example in a porous membrane or a chromatographic matrix, the transition in the macromolecule would affect the total free volume of the pores and hence present a means to regulate the porosity of the system. It has been shown possible to regulate the permeability of porous membranes by this mechanism. For example, certain pH- and thermosensitive polymers such as poly(methacrylic acid),[56] poly(benzyl glutamate), poly(2-ethylacrylic acid),[57] poly(4-vinylpyridine),[58] poly(N-acryloylpyrrolidine),[59] poly(N-n-propylacrylamide), or poly(acryloylpiperidine), poly(NIPAAM) alone,[29] or in copolymers with poly(methacrylic acid),[60] were grafted inside the pores of a suitable membrane and the construct was subsequently used as a thermo- or pH-sensitive chemical valve. Grafted poly(methacrylic) acid molecules will be charged at elevated pH and hence find themselves in an extended conformation. As a result the effective pore size of the membrane is smaller than the nominal one and for a given pressure the flow through the membrane is reduced ("off-state" of the membrane). With decreasing pH, the macromolecules become protonated, thereby losing their charge and concomitantly their hydrophilicity adopting a more compact conformation. The effective pore size and hence the flow through the membrane increases ("on-state" of the membrane). It has been shown that the flux of dextran molecules with molecular masses between 4,400 and 50,600 g/mol) across a temperature sensitive, poly(NIPAAM)-grafted membrane could be effectively controlled by the temperature, the ionic strength of the buffer and the graft density of the membrane, while the flux of small molecules, such as mannitol, was not affected by the temperature even for very high graft densities.[61] The on-off-permeability ratios for different molecules (water, chloride ions, choline, insulin and albumin) ranged between 3 and 10 and were generally found to increase with the molecular mass.[62] Polymeric composite films formed by multilayers of amine-terminated dendrimers and poly(maleic anhydride-comethylvinyl ether) grafted onto gold-coated silica exhibited a fully reversible, pH-dependent permselectivity for both cationic and anionic redox-active probe molecules.[63]

In the area of "smart" chromatographic supports, it is especially size exclusion chromatography (SEC) which is most likely to profit. SEC-supports have a well-defined pore size distribution. Depending on their own size, only a fraction of these pores may be accessible to the molecules in the sample mixture and separation generally depends on differences in the accessible pore volume. Since stimulus-responsive polymers allow reversible fine-tuning of the pore size of such supports, some interesting applications were suggested. In one example, poly(NIPAAM) oligomers with carboxyl functions as end groups were coupled to porous glass beads with pore diameters from 156 to 408 Å.[64] Between 25° C and 32° C there was an abrupt change in the elution behavior. Under otherwise identical conditions, the retention times (elution volumes) of the dextran molecules were consistently shorter at lower temperatures. Presumably, at lower temperature extended polymer chains impeded the penetration of the dextran molecules into the pores to some extent. The effect was largest for the glass beads with the smallest pore size (156 Å) and in this case, the change in the resolution of the mixture below and above the phase transition temperature was very pronounced. Apparently, the transition of the grafted macromolecules efficiently regulates the pore size and as a result also the elution profile of substances with different molecular masses. While the retention times increased with increasing temperature in case of the polymer-grafted glass beads, the opposite will be observed for stimulus-responsive hydrogels made from similar polymers. In the case of hydrogel beads composed of cross-linked poly(acrylamide-co-N-isopropylacrylamide) or porous polymer beads with grafted poly(NIPAAM), the components of the test mixture were observed to elute earlier as the temperature was raised, presumably due to the shrinkage of the pores.[65,66]

Especially in the area of chemical valves, a number of additional stimuli have been used to trigger the conformational change in the polymers. In one example, a copolymer of poly(NIPAAM)-triphenylmethane leucocianide was grafted to the pore surface and the resulting membrane was photosensitive. UV-irradiation was found to increase the permeation of diethylene glycol through the membrane.[67] A membrane changing its permeability in response to the presence of a particular substance was constructed using specific biorecognition elements, which could recognize the presence of these substances and translate the signal into a change of a physicochemical property, such as the pH. It was attempted to use this principle to create a stimulus-responsive "chemical valve", which releases insulin in response to the presence of glucose.[68] Glucose oxidase was immobilized on a pH-responsive, poly(acrylic acid)-grafted, porous polycarbonate membrane. At neutral conditions, the polymer chains are densely charged and have the usual extended conformation, thereby preventing insulin transport through the membrane (blocking of the pores). In the presence of glucose, the pH inside the membrane pores drops as a result of activity of the immobilized enzyme (glucose is oxidized by the glucose oxidase, releasing protons as a side product). At low pH, the polymers are protonated and become noncharged. The conformation becomes more compact, the pores are unblocked and insulin may pass the membrane (Fig. 6.4). Once the trigger (glucose) is removed, the membrane returns to its previous "off-state". The development of such systems is still in a very early state. However, the possible implications for the use of smart polymers in medical applications are obvious.

Systems like the one presented above could also be used for efficient drug delivery responding to the needs of the organism.[69] For example, a membrane consisting of poly(2-hydroxyethyl acrylate-co-N,N-diethylaminomethacrylate-co-4-trimethylsilylstyrene) was shown to contain 30% of water at pH 6.3 (collapsed or shrunken state), while the same membrane contains up to 53% of water at pH 6.15 (expanded or swollen state). The transition between the two states changes the membrane's permeability for insulin by a factor of 42. Capsules of this copolymer containing glucose oxidase and insulin, released insulin 5 times faster in the presence of 0.2 M glucose than when no glucose was present. When the glucose was removed, the insulin-release rate returned to the initial value.

Another way of controlling the porosity of a polymer network is by reversible cross-linking. For example, poly(m-acrylamidophenylboronic acid-co-vinylpyrrolidone) forms a gel in the presence of poly(vinyl alcohol) because of the strong interactions between its boronate groups with the hydroxy groups of the poly(vinyl alcohol). Any low molecular mass polyalcohol, for example also glucose, competes with poly(vinyl alcohol) for the boronate groups and this may eventually lead to the dissolution of the gel.[70] The boronate-poly(vinyl alcohol) system has, for instance, been used for constructing a glucose-sensitive system for insulin delivery.[71-73] If the gel is loaded with insulin, the glucose-induced transition of the polymer network from the gel to the sol state increased the release of insulin from the gel drastically, thus representing a system capable of controlled drug delivery in response to an external signal, in this particular case the presence of glucose. A reversible response to glucose has been implemented in another system, which uses concanavalin A, a protein containing four glucose binding sites, as a glucose

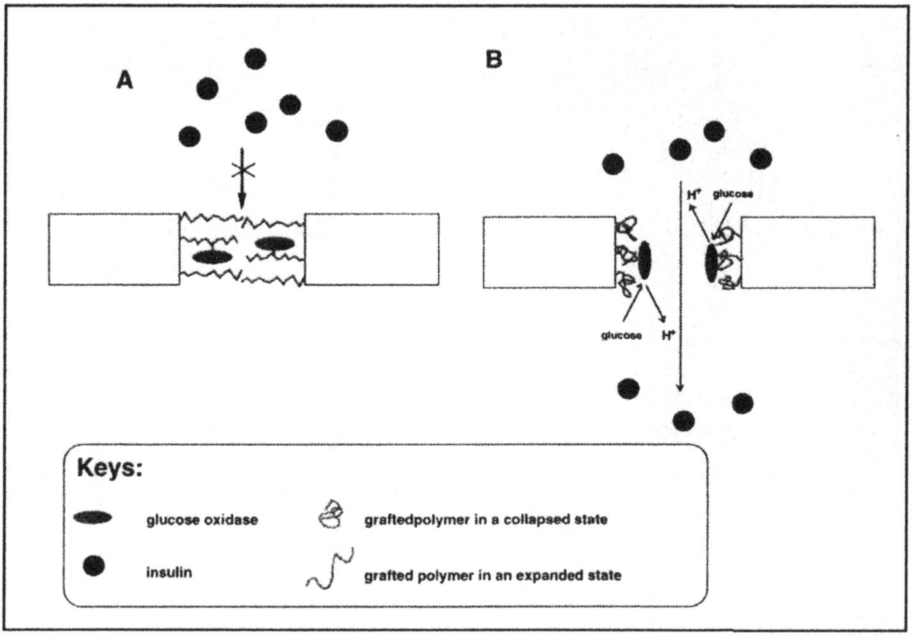

Figure 6.4. Schematic presentation of "chemical valve" with glucose oxidase immobilized on a pH-responsive polyacrylic acid grafted on a porous polycarbonate membrane: A—poly(acrylic acid) is in an expanded conformation blocking insulin transport; B—the oxidation of glucose is accompanied by decrease in pH and transition of poly(acrylic acid) into a compact conformation resulting in opening the pores for insulin transport (redrawn from [68]).

sensitive device.[74,75] Polymers with glucose molecules in the side chain such as poly(vinylpyrrolidone-coallylglucose) or poly(glucosyloxyethyl methacrylate), are reversibly cross-linked into a gel by concanavalin A. The addition of free glucose molecules results in displacement of the glucose-bearing polymer chains from the complex and consequently the dissolution of the gel.

Polymer-Carrying Liposomes for Triggered Release/Drug Delivery

The properties of lipid membranes (e.g., of liposomes) carrying a stimulus-responsive polymer may change in response to the phase transition of this polymer. The actual consequences of the conformational transition of the macromolecule on the liposome surface depend to a large extent on the fluidity and hence the composition of the membrane (Fig. 6.5).[76] If the membrane is in a fluid state at temperatures both above and below transition temperature of the polymer, the collapse of the polymer chains forces the anchor groups to move (by lateral diffusion) closer together. The result is the formation of compact globules of collapsed polymer attached to some isolated points on the membranes, while a major part of the liposome's surface is in fact not covered by the polymer. Such liposomes show little tendency to aggregate. However, if the membrane is in a solid state at temperatures both above and below the transition temperature of the polymer, the lateral diffusion of anchor groups is impossible and the collapsed polymer globules are found more or less evenly distributed over the liposome surface. The latter situation promotes significantly the aggregation and fusion of the liposomes which is often accompanied by the release of the liposome's content into the surrounding medium.[77]

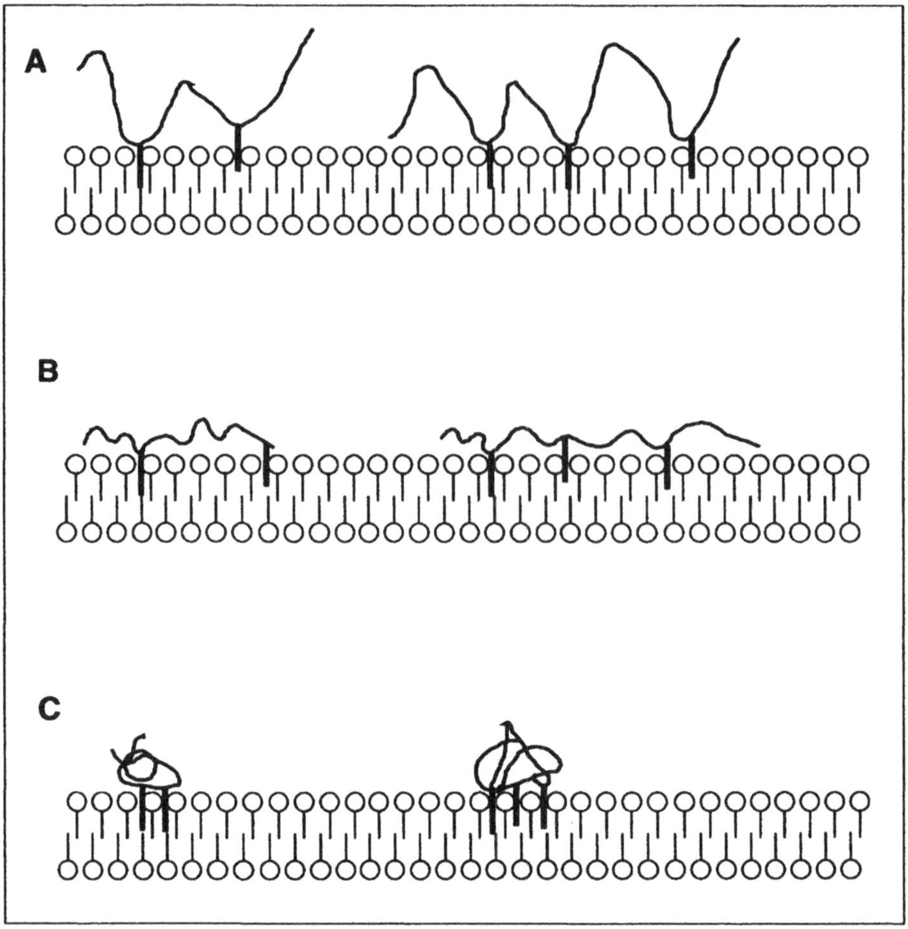

Figure 6.5. Schematic presentation of hydrophobically-modified smart polymers at the liposome surface. Expanded polymer chains stabilize liposomes against association at temperatures below polymer transition (A). At temperatures above polymer transition (B and C), the collapsed polymer chains cover nearly all the surface enhancing liposome association when liposome membrane is solid (B) or only small part of it exposing bare surface when liposome membrane is liquid (C).

The observation that the perturbation of the liposome membrane by the conformational transition of the attached polymer molecules affects both the aggregation tendency and their permeability for the substances incorporated in the liposome has been made for a number of systems. For example, poly(ethacrylic acid) undergoes a transition from the expanded to the compact form in the physiological pH-range of 7.4 to 6.5.[78] When covalently coupled to the surface of the liposomes formed from phosphatidylcholine, the pH-induced transition of the poly(ethacrylic acid) chains results in the fission of the large liposomes into more compact micelles with a concomitant release of the liposome's content into the surrounding environment. The temperature-induced transition of either poly(NIPAAM-co-N,N-didocecylacrylamide) or poly(NIPAAM-co-octadecylacrylate) was also found to enhance the release of the fluorescent marker, calcein, encapsulated in the copolymer-coated (phosphatidylcholine) liposomes.[79,80] On the other hand, the liposomes released almost no calcein at temperatures below the transition temperature of the polymer (near 32°C), whereas complete release was achieved within less than a minute at 40°C.

If the liposomes in question are treated with the polymer after their formation, the polymer binds only to the outer surface of the liposomes. If the liposomes are formed from a lipid-polymer mixture, on the other hand, the polymer is present on both sides of the liposome membrane. Such liposomes respond even faster to temperature changes.[31] The change of the liposome surface properties caused by the phase transition of stimulus-responsive polymers in also known to affect their interaction with cells. The phenomenon has been used in an attempt to develop a targeted drug delivery system. Liposomes modified with a pH-sensitive polymer, namely succinylated poly(glycidol), were shown to deliver the dye cacein more efficiently into cultured monkey kidney cells than nonmodified liposomes.[81]

An alternative mode to use smart polymers for targeted drug delivery was proposed in the form of polymeric nanoparticles with core-shell micelle structure prepared from an amphiphilic block-copolymer containing poly(NIPAAM) (thermo-responsive outer shell) by dialysis of the copolymer solution in dimethylformamide against water.[82] The polymeric micelles were very stable in aqueous solution and had a fairly long half-live (long blood circulation) because of their small diameter , their unimodal size distribution (24±4 nm) and their low critical micellar concentration of around 10 μg/mL. When the temperature was increased above the transition temperature of the thermo-responsive block chains (32° C), the outer shell chains dehydrated and collapsed, favoring the aggregation between micelles and concomitantly the interaction with cell membrane surfaces. The delivery of hydrophobic molecules contained within the micelles into the cell membranes was thereby facilitated. Such micelles may evolve into valuable agents for site-specific drug delivery using changes in temperature (such as the increased temperature in an inflamed tissue) as a trigger for the drug release.

Conclusions

The present chapter reviews some of the many new developments in the field of smart polymers. The review is not exhaustive—it is simply not possible to do this properly within the limits of a single chapter. We have tried to review some selected areas emphasizing mainly the application of the surfaces with grafted smart polymers. It seems appropriate to close this chapter with some hints and personal speculations about the future research avenues in the field. We look forward to further work in synthetic polymer chemistry which will provide new polymers capable of even stronger conformational transitions induced by changes in pH or temperature at a selected pH or temperature value. New techniques are expected for a more controlled modification of surfaces with polymers allowing the design of different architecture of polymer chains at the surface. On the theoretical side we hope to see considerable efforts towards a better understanding of the mechanism underlying the phase transition of polymers in aqueous solutions as well as a deeper insight into the colloid behavior of the polymer molecules at the solid-liquid interface.

References

1. Harland RS, Prud'homme RK, eds. Polyelectrolyte Gels. Properties, Preparation, and Applications. Washington DC: ACS, 1992:285-304.
2. Dusek K, ed. Responsive Gels: Volume Transitions II. Berlin: Springer-Verlag, 1993:180-197.
3. Dusek K, ed. Responsive Gels: Volume Transitions I. Berlin: Springer-Verlag, 1993:1-62.
4. Linné-Larsson E, Mattiasson B. Isolation of concanavalin A by affinity precipitation. Biotechnol Techniques 1994; 8:51-56.
5. Hoshino K, Yamasaki H, Chida C et al. Continuous simultaneous saccharification and fermentation of delignified rice straw by a combination of two reversibly soluble autoprecipitating enzymes and pentose-fermenting yeast cells. J Chem Eng Japan 1997; 30:30-37.

6. Park SY, Bae YH. Novel pH-sensitive polymers containing sulfonamide groups. Macromol Rapid Commun 1999; 20:269-273.
7. Baltes Th, Garret-Flaudy F, Freitag R. Investigation of the LCST of polyacrylamides in aqueous solution as a function of molecular parameters and solvent composition. J Polym Sci 1999; 37:2977-2989.
8. Heskins M, Guillet JE. Solution properties of poly(N-isopropylacrylamide). J Macromol Sci 1968:1441-1455.
9. Suwa K, Morishita K, Kishida A et al. Synthesis and functionalities of poly(N-vinylalkylamide). V. Control of a lower critical solution temperature of poly(N-vinylalkylamide). J Polymer Sci 1997; 35:3087-3094.
10. Tager AA, Safronov AP, Sharina SV et al. Thermodynamic study of poly(N-vinyl caprolactam) hydration at temperatures close to lower critical solution temperature. Colloid Polym Sci 1993:271.
11. Galaev IY, Mattiasson B. Thermoreactive water-soluble polymers, nonionic surfactants, and hydrogels as reagents in biotechnology. Enzyme Microb Technol 1993; 15:354-366.
12. Liu F, Liu FH, Zhuo RX et al. Development of a polymer-enzyme immunoassay method and its application. Biotechnol Appl Biochem 1995; 21:257-264.
13. Mori S, Nakata Y, Endo H. Purification of rabbit C-reactive protein by affinity precipitation with thermosensitive polymer. Protein Expression Purific 1994; 5:151-156.
14. Nonaka T, Yoda T, Kurihara S. Preparation of thermosensitive polyvinyl alcohol-graft-N-isopropylacrylamide copolymer membranes containing carboxyl groups and permeation of solutes through the membranes. Polym Mater Sci Eng 1994; 77:344.
15. Freitag R, Baltes TH, Eggert M. A comparison of thermoreactive water-soluble poly-N,N-diethylacrylamid prepared by anionic and group transfer poymerization. J Polym Sci 1994; 32:3019-3030.
16. Chen G, Hoffman AS. Graft copolymers that exhibit temperature-induced phase transition over a wide range of pH. Nature 1995; 373:49-52.
17. Winnik FM, Davidson AD, Hamer GH et al. Amphiphilic poly(N-isopropylacrylamide) prepared by using lipophilic radical initiator: Synthesis and solution properties in water. Macromolecules 1992; 25:1876-1880.
18. Schild HG. Poly(N-isopropylacrylamide): Experiment, theory and application. Prog Polym Sci 1992; 17:163-249.
19. Mattiasson B, Kumar A, Galaev IY. Affinity precipitation of proteins: Design criteria for an efficient polymer. J Molec Recogn 1998; 11:211-216.
20. Durand A, Hourdet D. Synthesis and thermo-associative properties in aqueous solution of copolymers containing poly(N-isopropylacrylamide) side chains. Polymer 1999; 40:4941-4951.
21. Yoshida R, Uchida K, Kaneko Y et al. Comb-like grafted hydrogels with rapid de-swelling response to temperature changes. Nature 1995; 374:240-242.
22. Yakushiji T, Sakai K, Kikuchi A et al. Effects of cross-linked structure on temperature-responsive hydrophobic interaction of poly(N-isopropylacrylamide) hydrogel-modified surfaces with steroids. Anal Chem 1999; 71:1125-1130.
23. Lee SJ, Park K. Synthesis and characterization of sol-gel phase reversible hydrogels sensitive to glucose. J Molec Recogn 1997; 9:549-557.
24. Wu KYA, Wisecarver KD. Cell immobilization using PVA crosslinked with boric acid. Biotechnol Bioeng 1992; 39:447-449.
25. Kokufuta E, Matsukawa S. Enzymatically induced reversible gel-sol transition of a synthetic polymer system. Macromolecules 1995; 28:3474-3475.
26. Takagi T, Takahashi K, Aizawa M et al. Proceedings of the First International Conference on Intelligent Materials. Lancaster: Technomic Publishing Co. Inc., 1993:383-388.
27. Yalpani M. Biomedical Functions and Biotechnology of Natural and Artificial Polymers. ATL Press, 1996:157-168.
28. Kubota H, Shiobara N. Photografting of N-isopropylacrylamide on cellulose and temperature-responsive character of the resulting grafted celluloses. Reactive Fuctional Polymers 1998; 37:219-224.
29. Lee YM, Shim JK Preparation of pH/temperature responsive polymer membrane by plasma polymerization and its riboflavin permeation. Polymer 1997; 38:1227-1232.

30. VonRecum HA, Kikuchi A, Okuhara M et al. Retinal pigmented epithelium cultures on thermally- responsive polymer porous substrate. J Biomater Sci 1998; 9:1241-1253.
31. Hayashi H, Kono K, Takagishi T. Temperature sensitization of liposomes using copolymers of N-isopropylacrylamide. Bioconjugate Chem 1999; 10:412-418.
32. Dubin PL, Tong P. Colloid-Polymer Intercations. Particulate, Amphiphilic, and Biological Surfaces. Washington DC: American Chemical Society, 1993:216-240.
33. Wu XS, Hoffman AS, Yager P. Conjugation of phosphatidylethanolamine to poly(N-isoropylacrylamide) for potential use in liposome drug delivery system. Polymer 1992; 33:4659-4662.
34. Kono K, Nakai R, Morimoto K et al. Thermosensitive polymer-modified liposomes that release contents around physiological temperature. Biochim Biophys Acta 1999; 1416:239-250.
35. Tirrell DA In Kost J. Pulsed and Self-Regulated Drug delivery. Boca Raton: CRC Press, 1990:110-116.
36. Liang L, Feng X, Liu J et al. Reversible surface properties of glass plate and capillary tube grafted by photopolymerization of N-isopropylacrylamide. Macromolecules 1998; 31:7845-7850.
37. Takei YG, Aoki T, Sanui K et al. Dynamic contact angle measurement of temperature-responsive surface properties for poly(N-isopropylacrylamide) grafted surfaces. Macromolecules 1994; 27:6163-6166.
38. Kondo T, Koyama,M, Kubota H et al. Characteristics of acrylic acid and N-isopropylacrylamide binary monomers-grafted polyethylene film synthesized by photografting. J Appl Polym Sci 1998; 67:2057-2064.
39. Yakushiji T, Sakai K, Kikuchi A et al. Graft architectural effects on thermo-responsive wettability changes of poly(N-isopropylacrylamide)-modified surfaces. Langmuir 1998; 14:4657-4662.
40. Schmitt FJ, Park C, Simon J et al. Direct surface force and contact angle measurements of an adsorbed polymer with a lower critical solution temperature. Langmuir 1998; 14:2838-2845.
41. Kanazawa H, Kashiwase Y, Yamamoto K et al. Temperature-responsive liquid chromatography. 2. Effects of hydrophobic groups in N-isopropylacrylamide copolymer-modified silica. Anal Chem 1997; 69:823-830.
42. Hosoya K, Kimata K, Araki T et al. Temperature-controlled high-performance liquid chromatography using a uniformly sized temperature-responsive polymer-based packing material. Anal Chem 1995; 67:1907-1911.
43. Go H, Sudo Y, Hosoya K et al. Effects of mobile-phase composition and temperature on the selectivity of poly(N-isopropylacrylamide)-bonded silica gel in reversed-phase liquid chromatography. Anal Chem 1998; 70:4086-4093.
44. Yakushiji T, Sakai K, Kikuchi A et al. Graft architectural effects on thermo-responsive wettability changes of poly(N-isopropylacrylamide)-modified surfaces. Langmuir 1998; 14:4657-4662.
45. Kanazawa H, Matsushima Y, Okano T. Temperature-responsive chromatography. Trends Anal Chem 1998; 17:435-440.
46. Yoshioka H, Mikami M, Nakai T et al. Preparation of poly(N-isopropylacrylamide)-grafted silica gel and its temperature-dependent interaction with proteins. Polym Adv Technol 1995; 6:418-420.
47. Millner PA. High Resolution Chromatography. IRL Press, 1996:153-190.
48. Galaev IY, Warrol C, Mattiasson B. Temperature-induced elution of proteins from dye-affinity columns using an immobilized polymeric displacer. J Chromatogr 1994; 684:37-43.
49. Absolom DR, Thomson C, Kruzyk W et al. Adhesion of hydrophilic particles (human erythrocytes) to polymer surfaces: effect of pH and ionic strength. Colloids Surf 1986; 21:447-456.
50. Yamada N, Okano T, Sakai H et al. Thermo-responsive polymeric surfaces: Control of attachment and detachment of cultured cells. Makromol Chem Rapid Commun 1990; 11:571-576.
51. Okano T, Yamada N, Okuhara M et al. Mechanism of cell detachment from temperature-modulated, hydrophylic/hydrophobic polymer surfaces. Biomaterials 1995; 16:297-303.
52. Yamato M, Okuhara M, Karikusa F et al. Signal transduction and cytoskeletal reorganization are required for cell detachment from cell culture surfaces grafted with a temperature-responsive polymer. J Biomed Mater Res 1999; 44:44-52.
53. VonRecum HA, Kim SW, Kikuchi A et al. Novel thermally reversible hydrogel as detachable cell culture substrate. J Biomed Mater Res 1998; 40:631-639.
54. Ito Y, Chen G, Guan Y et al. Patterned immobilization of thermo-responsive polymer. Langmuir 1997; 13:2756-2759.

55. Bohanon T, Elender G, Knoll W et al. Neural cell pattern formation on glass and oxidized silicon surfaces modified with poly(N-isopropylacrylamide). J Biomater Sci 1996; 8:19-39.
56. Ito Y, Ochiai Y, Park YS et al. pH-Sensitive gating by conformational change of a polypeptide brush grafted onto a porous polymer membrane. J Am Chem Soc 1997; 119:1619-1623.
57. Ito Y, Park YS, Imanishi Y. Imaging of pH-sensitive polymer brush on a porous membrane using atomic force microscopy in aqueous solution. Macromol Rapid Commun 1997; 18:221-224.
58. Okahata Y, Ozaki K, Seki T. pH-sensitive permeability control of polymer-grafted nylon capsule membranes. J Chem Soc 1984:519-521.
59. Okahata Y, Noguchi H, Seki T. Thermoselective permeation from a polymer-grafted capsule membrane. Macromolecules 1986; 19:493-494.
60. Nonaka T, Hashimoto K, Kurihara S. Preparation of thermosensitive cellophane-graft-N-isopropylacrylamide copolymer membranes and permeation of solutes through the membranes. J Appl Polym Sci 1997; 66:209-216.
61. Akerman S, Viinikka P, Svarfvar B et al. Drug permeation through a temperature-sensitive poly(N-isopropylacrylamide) grafted poly(vinylidene fluoride) membrane. Int J Pharm 1998; 164:29-36.
62. Spohr R, Reber N, Wolf A et al. Thermal control of drug release by a responsive ion track membrane observed by radio tracer flow dialysis. J Controlled Release 1998; 50:1-11.
63. Liu Y, Zhao M, Bergbreiter DE et al. pH-switchable, ultrathin perm-selective membranes prepared from multilayer polymer composites. J Am Chem Soc 1997; 119:8720-8721.
64. Gewehr M, Nakamura K, Ise N. Gel permeation chromatography using porous glass beads modified with temperature-responsive polymers. Makromol Chem 1992; 193:249-256.
65. Park TG, Hoffman AS. Estimation of temperature-dependent pore size in poly(N-isopropylacrylamide) hydrogel beads. Biotechnol Prog 1994; 10:82-86.
66. Hosoya K, Sawada E, Kimata K et al. In situ surface-selective modification of uniform size macroporous polymer particles with temperature-responsive poly-N-isopropylacrylamide. Macromolecules 1994; 27:3973-3976.
67. Kurihara S, Ueno Y, Nonaka T. Preparation of poly(vinyl alcohol)-graft-N-isopropylacrylamide binary membranes with triphenylmethane leucocyanide and permeation of solutes through membranes. J Appl Polym Sci 1998; 67:1931-1937.
68. Imanishi Y, Ito Y. Glucose-sensitive insulin-releasing molecular systems. Pure Appl Chem 1995; 67:2015-2021.
69. Ishihara K, Matsui K. Glucose-responsive insulin release from polymer capsule. J Polymer Sci 1986; 24:413-417.
70. Kitano S, Kataoka K, Koyama Y et al. Glucose-responsive complex formation between poly(vinyl alcohol) and poly(N-vinyl-2-pyrrolidone) with pendent phenylboronic acid moieties. Makromol Chem 1991; 12:227-233.
71. Kitano S, Koyama Y, Kataoka K et al. A novel drug delivery system utilizing a glucose responsive polymer complex between poly(vinyl alcohol) and poly(N-vinyl-2-pyrrolidone) with a phenylboronic acid moiety. J Controlled Release 1992; 19:162-170.
72. Shiino D, Kataoka K, Koyama Y et al. Controlled release of insulin from boronic acid gel under physiological conditions. J Controlled Release. 1994; 28:317-319.
73. Kataoka K, Miyazaki H, Miyazaki B et al. Totally synthetic polymer gels responding to external glucose concentration: Their preparation and application to on-off regulation of insulin release. J Amer Chem Soc 1998; 120:12694-12695.
74. Lee SJ, Park K. Synthesis and characterization of sol-gel phase reversible hydrogels sensitive to glucose. J Molec Recogn 1997; 9:549-557.
75. Nakamae K, Miyata T, Jikihara A et al. Formation of poly(glucosyloxyethyl methacrylate)-Concanavalin A complex and its glucose-sensitivity. J Biomater Sci 1994; 6:79-90.
76. Ringsdorf H, Venzmer J, Winnik FM. Interaction of hydrophobically modified poly(N-isopropylacrylamide) with model membranes—Or playing a molecular accordion. Angew Chem 1991; 30:315-318.
77. Hayashi H, Kono K, Takagishi T. Temperature-dependent associating property of liposomes modified with a thermosensitive polymer. Bioconjugate Chem 1998; 9:382-389.

78. Sugaj S, Nitta K, Ohno N et al. Conformational studies of poly(ethacrylic acid) in aqueous salts by potentiometric, viscometric, optical and [1]N-NMR measurements. Colloid Polym Chem 1983; 261:159-165.
79. Kono K, Henmi A, Yamashita H et al. Improvement of temperature-sensitivity of poly(N-isopropylacrylamide)-modified liposomes. J Controlled Release 1999; 59:63-75.
80. Hayashi H, Kono K, Takagishi T. Temperature-controlled release property of phospholipid vesicles dependent bearing a thermosensitive polymer. Biochim Biophys Acta 1996; 1280:127-134.
81. Kono K, Igawa T, Takagishi T. Cytoplasmic delivery of calcein mediated liposomes modified with a pH-sensitive g poly(ethylene glycol) derivative. Biochim Biophys Acta 1997; 1325:143-154.
82. Cammas S, Suzuki K, Sone C et al. Thermo-responsive polymer nanoparticles with a core-shell micelle structure as site-specific drug carriers. J Controlled Release 1997; 48:157-164.
83. Kumar A, Galaev IY, Mattiasson B. Metal chelate affinity precipitation: A new approach to protein purification. Bioseparation 1999; 7:185-194.
84. Pietruska N, Galaev IY, Kumar A et al. New polymers forming aqueous two-phase polymer systems. Biotechnol Progress 2000; 16:408-415.
85. Wang KL, Burban JH, Cussler EL. Hydrogels as separating agents. In: Dusek K, ed. Responsive Gels: Volume Transitions II. Berlin: Springer-Verlag, 1993:68-79.
86. Monji N, Hoffman AS. A novel immunoassay system and bioseparation process based on thermal phase separating polymers. Appl Biochem Biotechnol 1987; 14:107-120.
87. Tatsuma T, Saito K, Oyama N. Enzyme exchangeable enzyme electrodes employing a thermoshrinking redox gel. J Chem Soc 1994:1853-1854.
88. Markvicheva EA, Kuz'kina IF, Pashkin II et al. A novel technique for entrapment of hybridoma cells in synthetic thermally reversible polymers. Biotechnol Techniques 1991; 5:223-226.
89. Park TG, Hoffman AS. Immobilization of *Arthrobacter simplex* in a thermally reversible hydrogel: Effect of temperature cycling on steroid conversion. Biotechnol Bioeng 1990; 35:152-159.
90. Fujii M, Taniguchi M. Application of reversibly soluble polymers in bioprocessing. Trends Biotechnol 1991; 9:191-196.
91. Willner I, Rubin S, Zor T. Reversible light stimulated activation and deactivation of α-chymotrypsin by its immobilization in photoisomerizable copolymers. J Am Chem Soc 1993; 115:8690-8694.
92. Osada Y, Gong J. Stimuli-responsive polymer gels and their application to chemomechanical systems. Progr Polym Sci 1993; 18:187-226.

CHAPTER 7

Molecularly Imprinted Polymers:

A New Dimension in Analytical Bioseparation

Oliver Brüggemann

Introduction

Biological interactions, for example the interaction between an antibody and its antigen or between an enzyme and its substrate, are in most cases closely linked to the three-dimensional structure of the protein. Antibodies as well as enzymes are large (protein) molecules made up of long chains of amino acid whose sequence is predetermined by the genetic code. In addition, complex proteins may also contain sugar, lipid or phosphate groups. Proteins have many functions in living organisms. Enzymes are required for digesting food and thus supplying the organism with a usable form of energy. Antibodies constitute an important part of the molecular part of the immune system (humoral immune response). Bacteria, viruses and various toxins are marked as "foreign" by the means of such antibodies. The generation of antibodies seems to be almost unlimited in its diversity; in fact the immune system seems to be able to react with the formation of a suitable antibody to nearly every imaginable (natural and synthetic) antigen.

During the past century the secrets of the immune system and the immune response have been unraveled and the interaction between antibodies and antigens as well as their respective analogs has inspired the creation of many new techniques in biology, (analytical) chemistry and medicine. Scientists have applied this know-how for the development of new methods for vaccination, new analytical tools (e.g., immunoassays, immunoaffinity chromatography, both based on immobilized antibodies) or new approaches in synthesis and catalysis (for instance based on catalytic antibodies). But very soon they also experienced the limitations of these novel approaches. Nearly all the biological materials (ligands) used showed problems with stability, for example, towards heat, "extreme" pH values, organic solvents and digestion by microorganisms and proteolytic agents in general. While the specificity of the bio-components was unsurpassed, their very biological (i.e., degradable) nature prevented them from being used in many cases. Consequently, researchers thought about alternatives and found them in a technology termed "molecular imprinting", which retains the principles of "bio"-specific interactions but realizes them by purely chemical means.

The Principle of Molecular Imprinting

Decades ago, when the very nature of the immune response was still under discussion, Linus Pauling proposed the so-called instructive theory of antibody creation.[1] According to this theory antibody molecules containing flexible, functional groups were winding around the different antigens in question, thus capturing them in a specifically created interactive "pocket",

Synthetic Polymers for Biotechnology and Medicine, edited by Ruth Freitag.

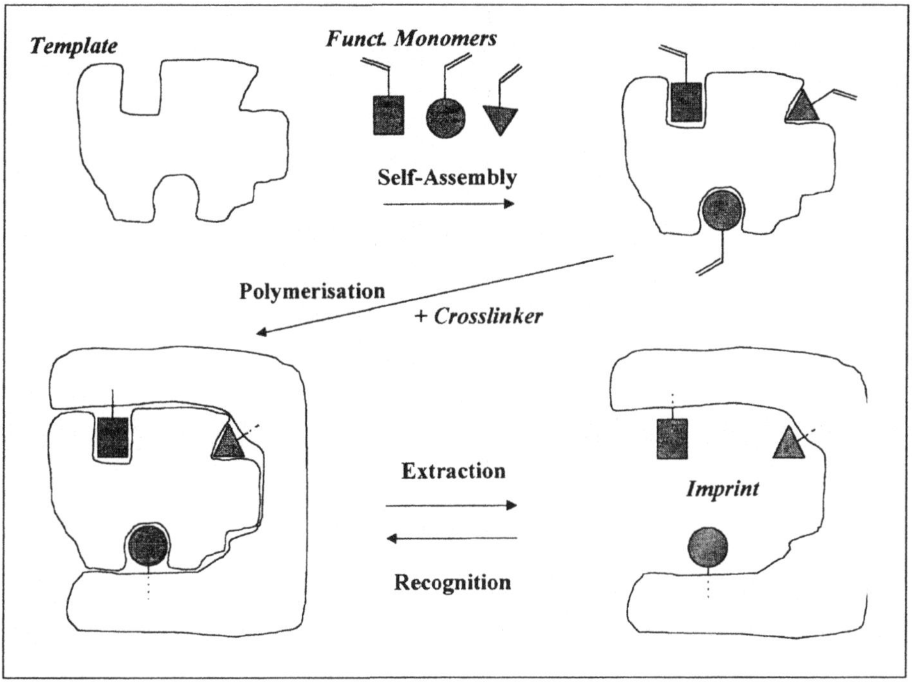

Figure 7.1. Principle of molecular imprinting.

where the interactive groups were arranged to correspond perfectly to those found on the surface of the antigen. The theory had to be abandoned later, when more details about the complex antigen-antibody-interaction became known. However, it was most instructive and the idea of an antigen being able to function as the template for the formation of a specific antibody (binding site) has in fact inspired the concept of molecular imprinting.

Molecular imprinting, MI, is a technique to generate polymeric mimics of biological receptors, for example antibodies or enzymes. In a first step a template (the antigen or substrate) is associated with functional monomers, i.e., molecules containing both a point of interaction (covalent or noncovalent bond) and a polymerizable group (usually a double bond). In a second step the polymerization is initiated in the presence of high amounts of cross-linking molecules containing at least two polymerizable groups (Fig. 7.1). These cross-linkers are necessary to establish a rigid polymer network, which maintains imprints with high specificity. After an extraction step to remove the template, the produced "Molecularly Imprinted Polymer" (MIP) is able to recognize specifically the template due to selective interaction of this molecule with the cavities created in the polymer network.

In the 1970's Wulff et al developed an imprinting method based on the covalent immobilization of cis-diol-sugar templates via vinylized boronate ester bridges to the functional monomer (Fig. 7.2a).[2] The template molecules were extracted from the polymer networks by hydrolysis. The remaining MIP cavities showed high specificity for the template sugars. The disadvantage of this procedure was the time-consuming step of derivatizing the templates prior to the polymerization and the slow process of reaching equilibrium during the application of the MIP for bioseparation. The latter was due to the necessity of the templates to form covalent bonds to the polymer for recognition.

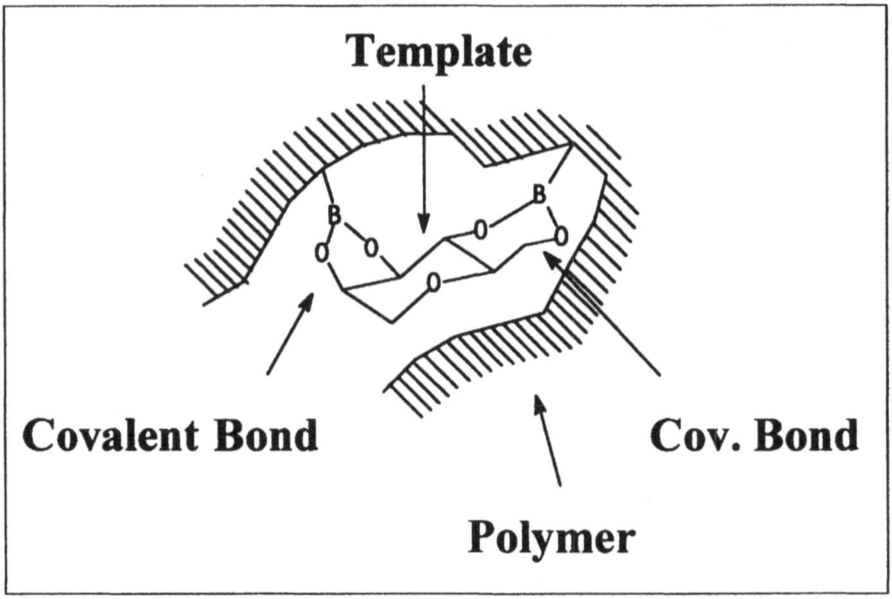

Figure 7.2a. Covalent molecular imprinting.

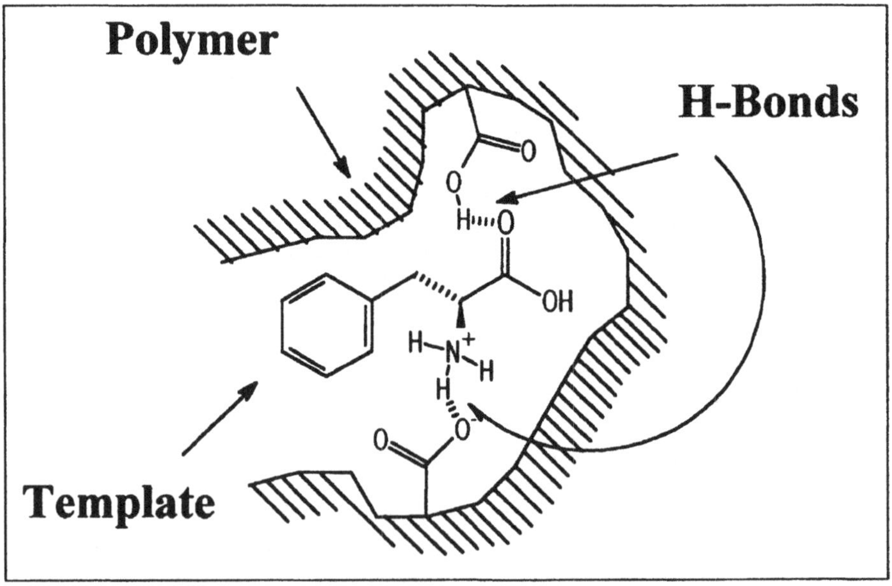

Figure 7.2b. Noncovalent molecular imprinting.

Mosbach et al pioneered a noncovalent way to generate MIP, which just calls for a self-assembly of template and functional monomer (Fig. 7.2b). The polymerizable "functional monomers" may associate with the template via a variety of interactions, including H-bridge formation, electrostatic and hydrophobic interaction. The polymerizable groups of the functional monomers are again incorporated in a highly cross-linked polymer network and hence once more a specific imprint (cavity) is created in the polymer.[3] After template extraction under relatively gentle conditions, the noncovalent MIP have been successfully used in areas such as separation techniques, catalysis and sensor technology. Since only noncovalent interactions are operative during both imprinting and application of the MIP, the adsorption kinetics are usually much faster than in case of the covalent ones.

Durability and stability are the main advantages of biomimetic polymers over true biological ligands, particularly in harsh environments like organic solvents, which are often used in analytical or catalytic applications. It has been shown that MIP can be used for months without any loss of efficiency and, in contrast to sensitive biological materials, can be stored and cleaned without taking excessive precautions. On the other hand, MIP have shown applicability similar to their biological counterparts as highly selective affinity stationary phases in analytical chemistry, as substitutes for antibodies in immunoassays, as enzyme-like catalysts or as recognition element for "bio"sensor development.[4-6]

For the production of an MIP, one has to take into account five different components: the template, the functional monomer, the cross-linker, the porogen and the initiator. The latter two determine the morphology of the polymer. The template and its functionalities usually determine the choice of the functional monomer. In noncovalent MI, the affinity between the template and the functional monomer should be as high as possible. For templates carrying amino groups, acidic monomers such as acrylic or methacrylic acid, have given good results. For templates with carboxylic acid groups, the best choice seems to be a basic functional monomer, for example vinylpyridine, as counterpart. These monomers will lead to ionic interactions, which are stronger than hydrogen bonds or just hydrophobic ones. Table 7.1 lists a number of functional monomers which are suitable for noncovalent imprinting. Different types of interaction are possible, including ionic and/or hydrogen bonds, electrostatic, hydrophobic/hydrophobic and/or van der Waals interactions.

An alternative for obtaining strong interactions is the use of metal ions for the generation of metal chelate complexes.[7] The metal, for example copper, acts as a bridge between the actual functional monomer and the template. An example is the interaction between 4(5)-vinylimidazol as functional monomer and an amino acid as template in the presence of copper ions. The copper coordinates the two imidazole nitrogens of the monomer and the amino and the carboxyl group of the template respectively. By applying this technique it is possible to even produce enantioselective cavities. The specificity of a given imprint is mostly determined by the strength of the template monomer interaction, generally speaking, the higher the affinity the better. Additionally, the number of interaction points between the template/analyte and the MIP surface should be as high as possible. Theoretically at least, three interaction points are necessary to determine a three dimensional structure of the template molecule within the imprint.

In case of covalent interactions the choice of the functional monomer, which has to be linked covalently to the template prior to imprinting, depends on the chemistry of the template. Templates containing cis-diol groups are usually esterified with a vinyl-boronic acid component to yield a reversible bis-ester. If this is not possible, alternatives are the generation of Schiff-bases (azomethines). This requires an amino group on the template and a carboxyl acid group on the functional monomer or vice versa (Table 7.2).

The cross-linker provides up to 90% of the polymerizable groups and thus determines the "polymer chemistry" of the MIP. Amongst other things it acts as a solvent for the template and

Table 7.1. Functional monomers for use in noncovalent molecular imprinting

1-Vinylimidazole 	4(5)-Vinylimidazole 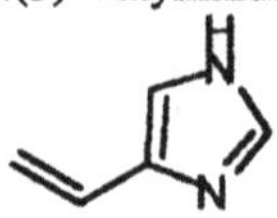	2-Vinylpyridine 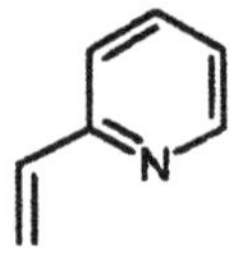
Acrylic acid 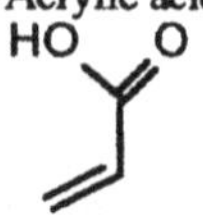	Methacrylic acid 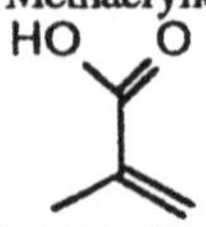	Methyl methacrylate 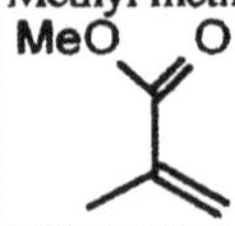
Itaconic acid 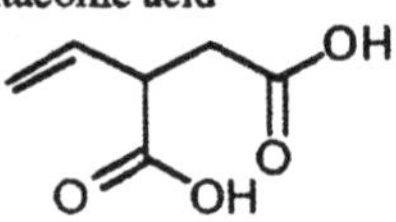	p-Vinylbenzoic acid 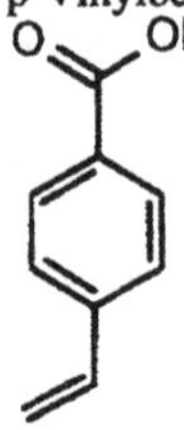	Ethylstyrene 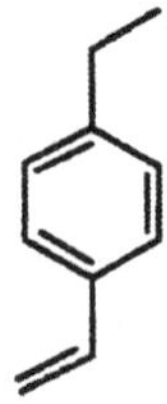
(2-Methacrylamido)-6-methylpyridine 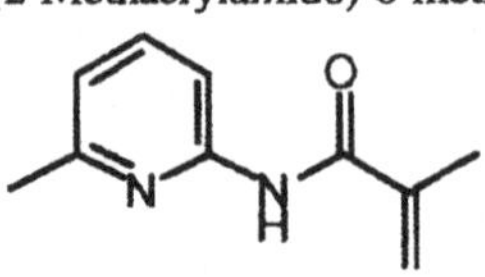	2-Acrylamido-2-methyl-1-propane-sulfonic acid 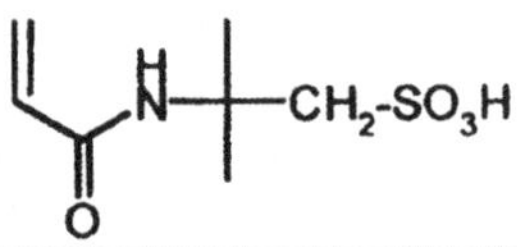	

the porogen (see below). The second related and equally important question concerns the number of polymerizable groups per cross-linker molecule. The typical cross-linkers used for the production of MIP are ethyleneglycol dimethacrylate (EGDMA, in some publications abbreviated EGDM or EDMA) and divinylbenzene (DVB), both containing only two polymerizable vinyl groups. Recently it has been shown that cross-linkers with more than two active groups, for example trimethylolpropane trimethacrylate (TRIM), lead to more porous materials with presumably improved mass transfer properties (better access to the imprinted cavities), hence leading to a higher capacity of the MIP. Table 7.3 gives an overview of cross-linkers commonly used for in MI experiments.

Most MIP today are prepared via standard free radical polymerization in dispersion. The initiator for the radical polymerization has to be selected with regard to its reliability. There are a few azo-bis compounds like AIBN[1], which are frequently used in thermo-initiated polymerization, but for UV-initiated processes, ABCHC seems to be more reliable than AIBN. For some special approaches, where molecular imprinting takes place in aqueous media, a water-soluble initiator like ABDV has to be chosen.

The porogen (solvent) is finally responsible for the formation of pores within the polymer network (Fig. 7.3).[8] Under the conditions of a typical dispersion polymerization, the polymer precipitates as it forms, still containing high amounts of porogen within the pores. For a successful imprinting, the template should be soluble in the porogen (sometimes the mixture has to be heated to assure this), but the porogen should not compete with the functional monomer for interaction with the template. If this interaction is, for example, mainly to be based on

Table 7.2. Functional monomers for use in covalent molecular imprinting

p-Vinylphenylboronic acid (forming boronates with cis-diol-templates) 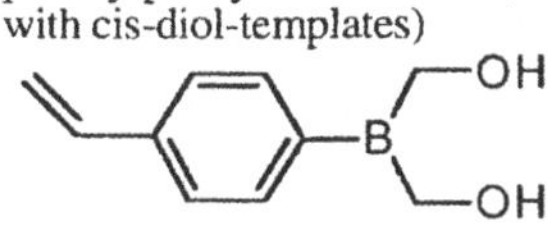	p-Vinylbenzyl-amine (forming azomethines and amides with carbonyl- and carboxyl-templates) 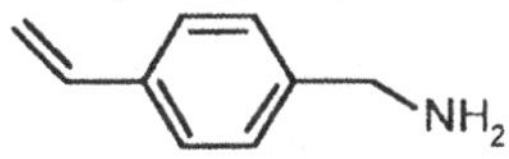
2-(p-Vinylphenyl)-1,3-propane diol (forming ketales with carbonyl-templates) 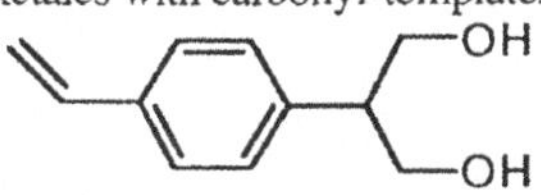	p-Vinylbenzyl alcohol (forming ether and ester with alcohol- and acid-templates) 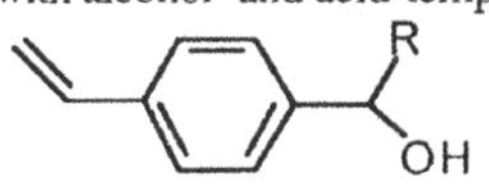
p-Vinylphenyl carbonyl (forming Schiff bases with amine-templates) 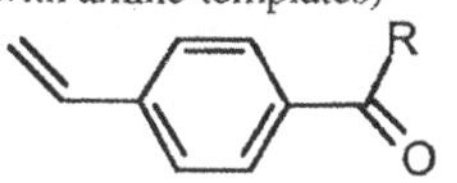	p-Vinylbenzoic acid (forming ester and amides with alcohol- and amine-templates respectively) 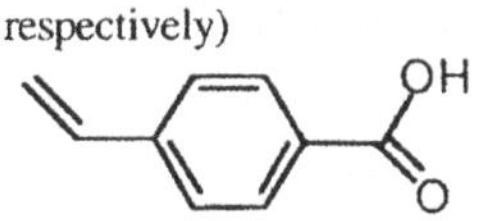

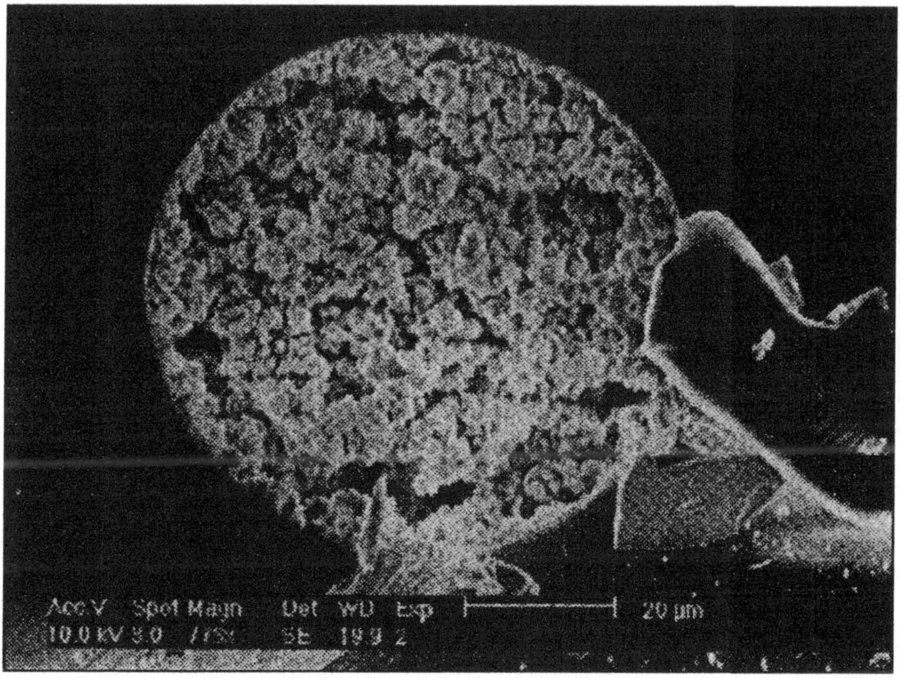

Figure 7.3. Scanning electron microscopic picture of a MIP inside a silica capillary.

hydrogen bonds, it seems obvious that a protic porogen would interfere and inhibit at least part of the template to interact with the functional monomers. In such cases an aprotic and apolar porogen should be used. In addition, the reaction mixture should be absolutely water-free, because water will very efficiently compete for hydrogen bonds. Dissolved oxygen has to be

Table 7.3. Cross-linkers for use in molecular imprinting

EGDMA 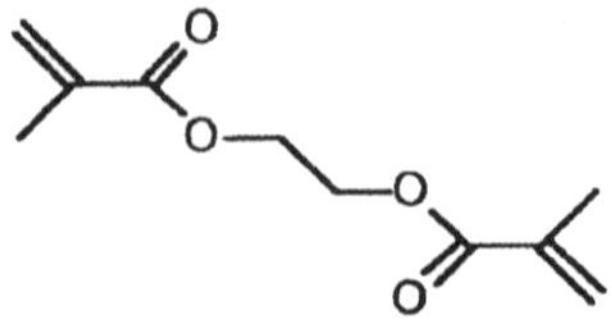	p-DVB 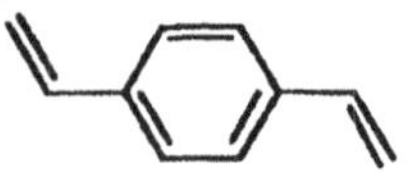
N,N'-methylendiacrylamide 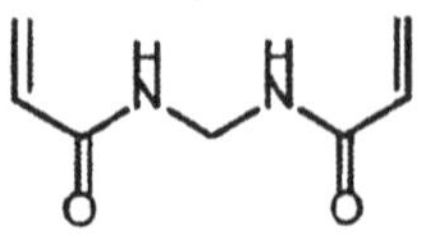	N,N'-1,4-phenylendiacrylamide 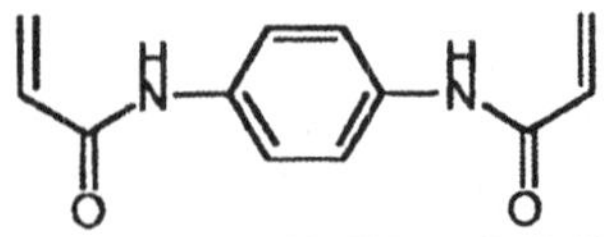
N,O-bisacryloyl-L-phenylalaninol 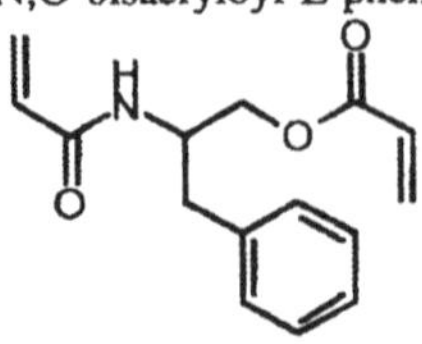	Trimethylolpropane trimethacrylate (TRIM) 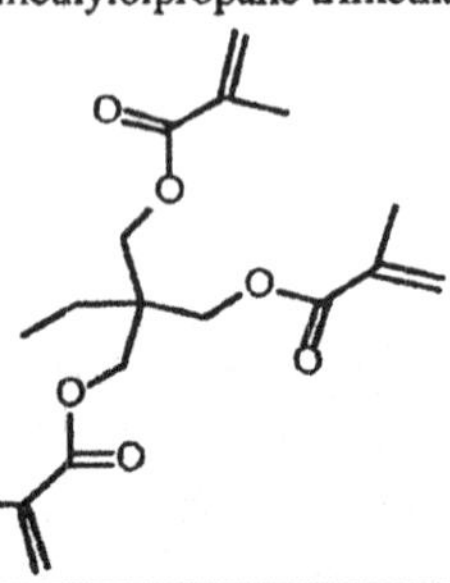
Pentaerythritol triacrylate 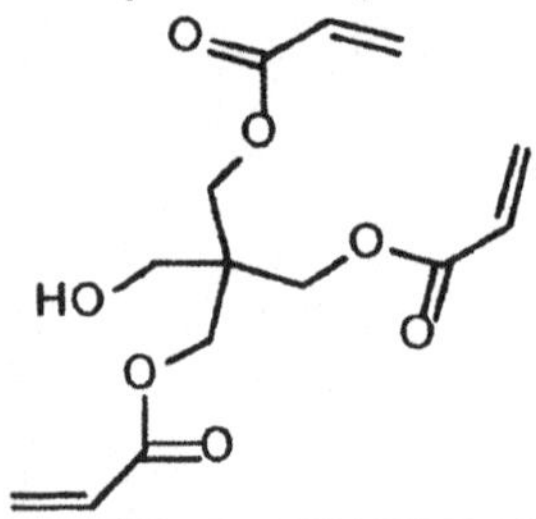	Pentaerythritol tetraacrylate 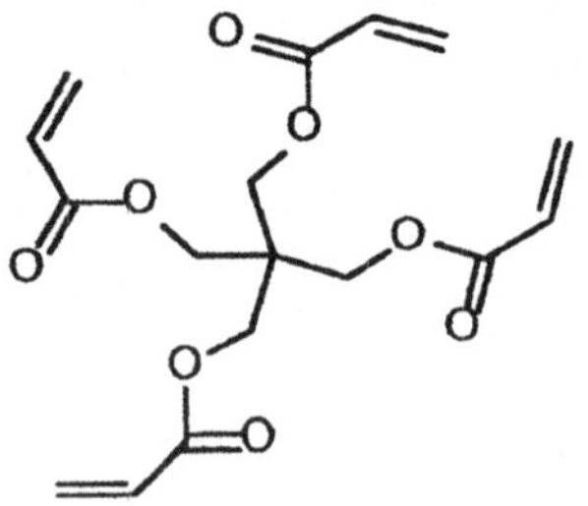

removed from the porogen via purging with nitrogen, because of the oxygen's tendency to inhibit radical polymerization in general. Commonly used porogens are DMF, DMSO, ACN, THF and chloroform. For more specific conditions, the reader is referred to the many applications cited below.

The following general protocol can be given for the preparation of MIP (and the control polymers prepared in a similar manner but in the absence of the template).

1. Dissolution of the template in the porogen, either at room temperature or at elevated temperature (maximum: temperature during polymerization).
2. Addition of the functional monomer and the cross-linker (in some cases this step also heLPS to dissolve remaining undissolved template molecules, due to complexation).
3. Addition of the initiator followed by purging of the mixture with nitrogen in order to remove the polymerization inhibitor oxygen.
4. Light (UV) or temperature induced polymerization in a UV-reactor, water-bath or oven at a defined temperature from –20° C to +65° C.

The polymers are generated in different configurations, mostly as bulk polymers, sometimes also as "monosized" beads, films or membranes. For some applications the polymers have to be processed further (for example, grinding or sieving in the case of bulk polymers). However, beads with a defined size or a very narrow range of diameters can be utilized directly without any further processing.

MIP for Bioseparation

Applications of MIP can be found in many areas of analytical (bio-)chemistry, such as chromatography, electrochromatography or membrane technology. On the other side, many different substances have been used as templates, including carbohydrates, amino acids and peptides/proteins, as well as pesticides and drugs. A noncomprehensive list of templates is given in Table 7.4.

HPLC—Affinity Liquid Chromatography

Molecularly imprinted polymers have been used in (affinity) chromatography from the very beginning. Before applying MIP in HPLC the following procedure usually has to be performed. After polymerization, the bulk polymers, which should still contain the template, are ground into pieces, wet sieved (e.g., using acetone and a 25 μm mesh sieve) and sedimented several times to remove fine particles ("fines"), which would otherwise threaten to clog the HPLC column. After drying the polymers are suspended in the packing solvent (e.g., chloroform or acetone), sonicated and slurry packed into a HPLC column under high pressure. After that packing procedure the template is extracted from the MIP with acidic or basic solvents (strong competitors for the binding). In order to assure comparability both the MIP and any putative control polymer should subjected equally to all procedures (swelling effects etc.), this includes steps like template elution and equilibrating the stationary with the mobile phase. Finally, a sample, dissolved preferably in mobile phase, can be injected. Typically, any substance with is identical or analogous to the original template molecule can be expected to interact strongly with the MI stationary phase and to elute with the highest retardation. Many applications show extremely high selectivities of these MI affinity phases, even allowing for efficient enantioselective separations. Unfortunately, the template peak commonly shows a pronounced broadness (tailing), caused by a certain (undesired) heterogeneity within the quality of the imprints.

Kempe used the MI technique to generate stationary phases imprinted with the peptide (Z)-L-Ala-L-Ala-OMe, using MAA and TRIM as monomers as well as chloroform as porogen.[9] Using mixtures of chloroform and acetic acid as mobile phases, it was possible to separate (Z)-D-Ala-D-Ala-OMe from (Z)-L-Ala-L-Ala-OMe under isocratic conditions obtaining a selectivity of 1.92 (Fig. 7.4b). The template isomer eluted in the second, broad peak, which is typical when using MI stationary phases, especially in combination with isocratic elution. Subsequently it was shown that gradient elution leads to sharper peaks for both isomers including the template species (Fig. 7.4a) and an improved selectivity of 3.19. Kempe et al also demonstrated the chirality of a polymer imprinted with Cbz-L-Tyr-OH by separating a Cbz-Tyr-OH racemate with a separation factor of 2.86.[10] Yu et al were able to obtain a separation factor of 2.69 when separating a racemic Boc-Trp-OH mixture on a Boc-L-Trp-OH imprinted stationary phase in a gradient of ACN and AcOH.[11] A baseline separation of the enantiomers was possible and due to the application of a gradient, the second (template) peak was less broad than in similar (isocratic) elution experiments. Another illustrative example for the resolution power of MI stationary phases is the use of polymers imprinted with oxacillin, a β-lactam-antibiotic. Oxacillin was imprinted using 4-Vinylpyridine as functional monomer and TRIM as cross-linker in ACN as porogen.[21] The resulting MI stationary phases were

Table 7.4. Important print molecules used for generating MIP applied in separation techniques

Class of Compounds	Template	Reference
Amino acids, peptides, proteins	Cbz-L-Glu-OH	9
	(Z)-L-Ala-L-Ala-OMe	10
	Boc-L-Trp-OH	11
	L-phenylalanine	12
	Boc-L-Trp	13-18
	Boc-L-Phe	19
	RNase A	20
Drugs	Oxacillin	21
	R-propranolol	22-25
	Theophylline	26-30
	Pentamidine	31
Pesticides, herbicides	S-(+)-2-phenylpropionic acid	32
	Atrazine	33-35
	2,4-Dichlorophenoxyacetic acid	36
Nucleotides	NAD^+	37
Nucleotide bases	9-Ethyladenin	38
Carbohydrates	Galactose derivatives	39
	Fructose derivatives	40
Steroids	Cholesterol	41
Metal ions	Cu^{2+}	7
Dyes	Rhodanile blue	42

evaluated both with organic and with aqueous mobile phases. Figure 7.5a shows the analysis of a mixture of oxacillin, penicillin V, penicillin G and a non-β-lactam-antibiotic (bacitracin). Oxacillin was baseline separated from the other penicillins (V and G), although these three substances are rather similar in structure. Using the control polymer for this analysis (Figure 7.5b) under the same conditions did not lead to a separation of oxacillin from the penicillins. Both the MI and the control column allow the separation of the β-lactam antibiotics from the non-β-lactam antibiotic bacitracin. Apparently the structures are sufficiently different in this case to allow separation by purely chromatographic means.

An alternative type of molecular imprinting has been proposed by Mayes et al.[19] Instead of generating bulk polymers, which usually entails a loss of up to 50% of the material during the subsequent time-consuming sedimentation steps, they used a suspension polymerization technique to directly fabricate MIP beads. The most important factor in this case was the use of a liquid perfluorocarbon (perfluoro(methylcyclohexane)) as the dispersing phase, which allowed the formation of stable emulsion droplets of the imprinting mixture (template, functional monomers, cross-linker, initiator and a porogen), in combination with the use of fluorinated surfactants and surface-active polymers as stabilizers. After polymerization, spherical beads of controllable size (distribution) were obtained, depending on the quantity of stabilizing polymer. When beads were produced according to a standard recipe (2 g monomer in 5 ml volume) using e.g., 25 mg of perfluoro polymeric surfactant (PFPS) a mean particle size of 19.7 μm was obtained with a standard deviation of 0.6 μm (Fig. 7.6). Particles could be manufactured with

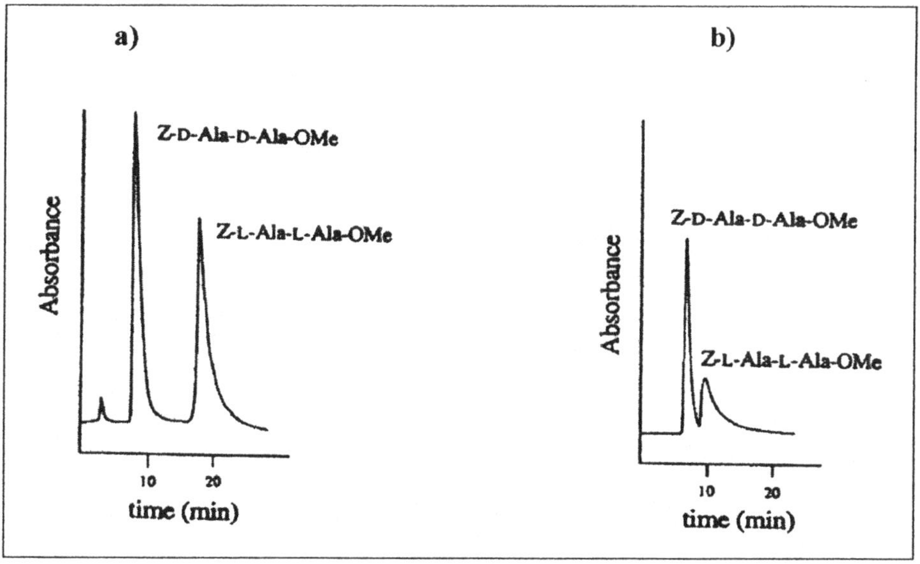

Figure 7.4. Chromatographic resolution of (a) 100 μg of a mixture of (Z)-L-Ala-L-Ala-OMe and (Z)-D-Ala-D-Ala-OMe on a (Z)-L-Ala-L-Ala-OMe-imprinted MIP using gradient elution; (b) 1 mg of a mixture of (Z)-L-Ala-L-Ala-OMe and (Z)-D-Ala-D-Ala-OMe on a (Z)-L-Ala-L-Ala-OMe-imprinted MIP using isocratic elution. Reprinted from: Kempe M, Mosbach K. Receptor binding mimetics: A novel molecularly imprinted polymer. Tetrahedron Letters 1995; 36:3563-3566. With permission from Elsevier Science.

sizes in a range from 5 μm to 50 μm. After a single washing step these beads were ready to be packed into HPLC columns and to be used for pseudo-affinity chromatography. MIP particles imprinted with Boc-L-Phe have been applied for chiral separation of racemic Boc-DL-Phe mixtures, which resulted in near baseline separation of the two enantiomers. The chromatogram of this racemate showed the typical broad peak for the template molecule. A separation factor of 1.91 was obtained using chloroform with 0.25% acetic acid as mobile phase.

HPCE—Affinity Capillary Electrochromatography

Capillary electrochromatography (CEC)—in particular in combination with molecular imprinting[43]—is a relatively new technique in the field of capillary electrophoresis (CE). CE separations in general take place in a thin (I.D. 25 μm to 100 μm), buffer-filled (silica) capillary. Both ends of the capillary are immersed in buffer vials, which also are connected via electrodes to a power supply. This allows the application of an electrical field (several kV) across the capillary thus separating charged substances according to their mass-to-charge ratio by electrophoretic and electro-osmotic effects. In CEC, the capillary contains an interactive stationary phase together with the electrophoresis buffer. As a result the separation is in addition to electrophoresis and electroosmosis also governed by chromatographic interactions between the stationary phase surface and the analytes. In many CEC-applications the mobile phase is driven not only by electroosmosis but also by the application of a slight pressure.

For pseudo-affinity CEC with MI stationary phases, capillaries have been filled with suspended MIP particles, e.g., as an additive to a gel-filling. Packing thin capillaries with HPLC-type stationary phases (particle diameter between 2 μm and 20 μm) is relatively difficult, so an alternative was sought. Proposed was, for example, the in situ polymerization of a suitable

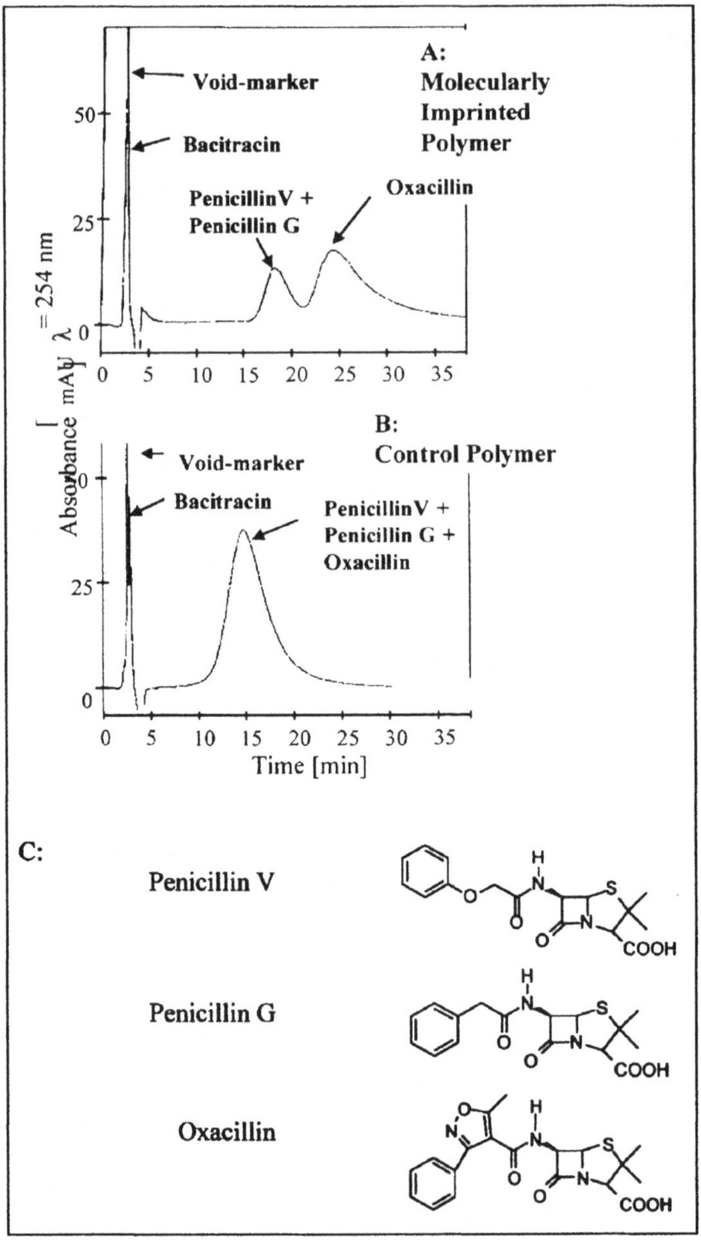

Figure 7.5. A: Chromatogram of a mixture containing the print molecule (oxacillin), two other β-lactam-antibiotics (penicillin G and penicillin V) and a non-β-lactam-antibiotic (bacitracin) on an oxacillin imprinted MIP containing 4-vinylpyridine residues, cross-linked with TRIM. The analysis was performed in organic mobile phase (ACN/AcOH, 99:1), B: same conditions but using the respective control polymer, C: Structures of penicillin V, penicillin G and oxacillin. Reprinted with permission from: Skudar K, Brüggemann O, Wittelsberger A et al. Selective recognition and separation of β-lactam antibiotics using molecularly imprinted polymers. Anal Commun 1999; 36:327-331.

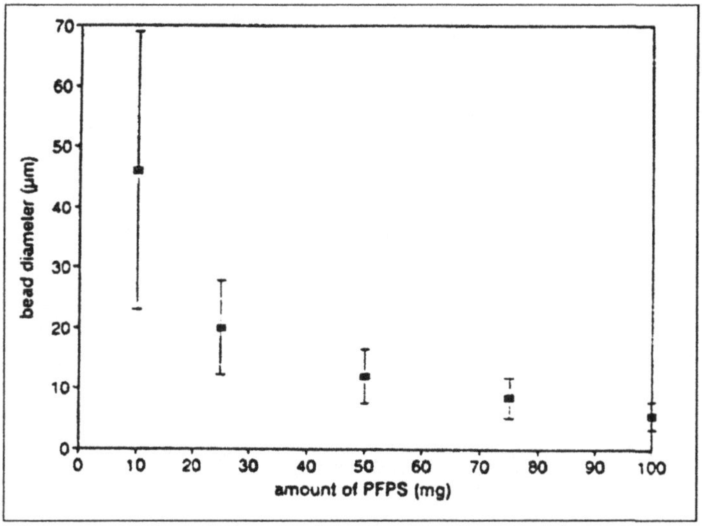

Figure 7.6. Bead diameter as a function of the amount of the stabilizing polymer (PFPS) added during a "standard polymerization" containing 1.84 g EGDMA, 0.16 g of MAA, 4.2 g of chloroform, and 20 mg of AIBN emulsified in 20 ml of perfluoro(methylcyclohexane). Reprinted with permission from: Mayes AG, Mosbach K. Molecularly imprinted polymer beads: Suspension polymerization using a liquid perfluorocarbon as the dispersing phase. Anal Chem 1996; 68:3769-3774.

stationary phase (wall-coating[32], macroporous monolith[22,23]) within the capillaries; an approach, which can easily be combined with MI. In the cited cases, the complete imprinting mixture was usually flushed into the capillary and polymerized at high temperatures in an oven or via UV-initiation at low temperature. After extracting the template from the polymer the capillaries were ready for use. To date, such MI capillaries have been mainly used for the separations of racemic mixtures in capillaries imprinted with one of the pure enantiomers in question.

When capillaries with an MI inner wall coating were investigated, the imprinting effect was not very pronounced and further optimization of this approach is clearly necessary.[32] For the separation of a racemic mixture of 2-phenylpropionic acid (Fig. 7.7a) a capillary coated with an S(+)-2-phenylpropionic acid imprinted MIP was used. Only one peak was observed, which was identified as the R(-)- enantiomer not used during imprinting. The electropherogram of only the R(-)- enantiomer (Fig. 7.7b) shows a similar peak. When the imprinting S(+)-2-phenylpropionic acid enantiomer was analyzed (Fig. 7.7c) a very broad peak (starting after 15.5 min) was observed, due to a very strong interactions with the MI stationary phase.

The approach proposed by Schweitz, Nilsson and coworkers seems to be more promising for pseudoaffinity CEC. This approach establishes a continuous, macroporous MI stationary phase within the capillary. In the usual protocol, a mixture of MAA (functional monomer), TRIM (cross-linker) and the template (e.g., R-propranolol), is polymerized in situ by UV-initiated radical polymerization at a temperature of –20°C. The resulting capillaries have been used in enantioselective CEC for the separation of rac-propranolol. Baseline separations of the enantiomeric analytes could be demonstrated (Fig. 7.8) as well as separations of structurally related racemates.[22,23]

Lin et al used MIP-type stationary phases embedded into a simple polyacrylamide gel for chiral CEC-separations.[12] L-phenylalanine imprinted MAA-EGDMA-polymer particles were able to achieve a baseline resolution of the corresponding D- and L-enantiomers, with the template molecule again occurring as the second, broader peak. The authors used a particle

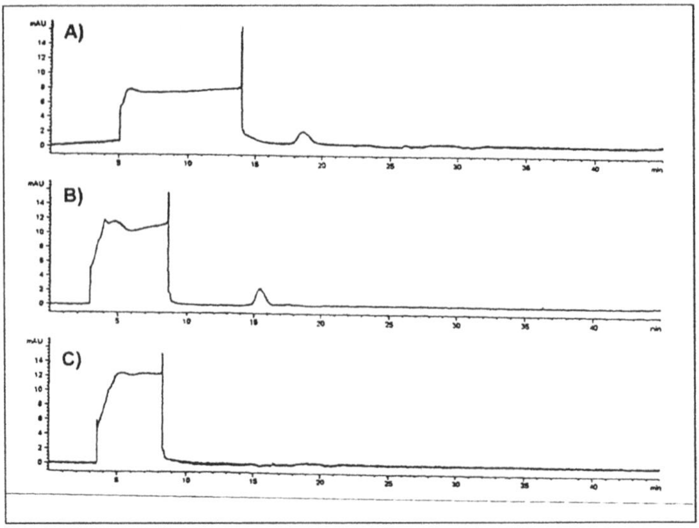

Figure 7.7. Electropherograms of (A) a racemic mixture of 2-phenylpropionic acid, (B) R(-)-2-phenylpropionic acid and (C) S(+)-2-phenylpropionic acid (using in all three experiments the MI capillary prepared using S(+)-2-phenylpropionic acid as template). Reprinted from Brüggemann O, Freitag R, Whitcombe MJ et al. Comparison of polymer coatings of capillaries for capillary electrophoresis with respect to their applicability to molecular imprinting and electrochromatography. J Chromatogr 1997; 781:43-53. With permission from Elsevier Science.

concentration of 10 mg/l and observed a dependence of the resolution on the particle size. In particular, a diameter of ≤ 5 μm gave the highest achievable resolution of 1.45 in this case.

Walshe et al[24] were able to resolve enantiomer mixtures using MIP as chiral additives in a background electrolyte at a level of 0.05%. The MIP was imprinted with S-propranolol as template, using N-acryloylalanine and EGDMA as monomers. Interestingly, MAA was not suitable as functional monomer in this case. The MIP was prepared as a bulk polymer and subsequently ground, sieved and finally extracted. Particles suspended in buffer were 20 μm to 30 μm in size. Figure 7.9 shows excellent baseline separation of the racemic mixture with narrow peaks being observed for both analytes. Separation of enantiomers seemed to be due to a difference in the electrophoretic mobility between the MIP-complexed and the free enantiomers, with the enantiomer used as print molecule having a much higher affinity of the MIP than the other one.[24]

MI-Membranes

Molecularly imprinted membranes were first described in 1995.[33] The idea behind the preparation was similar to the one, which led to the combination of MI and HPLC. Membranes with highly specific affinities towards one type of molecule can be used as "affinity filters" for a number of analytical and semi-preparative purposes. Piletsky et al were among the first, manufacturing a sensor membrane imprinted with atrazine.[33,34] For the actual production of the membrane they used the typical reaction mixture of template/MAA/EGDMA/ AIBN (initiator) and DMF (porogen) and polymerized the mixture on a glass filter surface. After extracting the template, the membrane showed a quantitative change in electrical resistance whenever getting back into contact with atrazine. The membranes showed a response to atrazine in the range of 0.01 mg/ml to 0.5 mg/ml using conductimetric detection. Sergeyeva et al have also described atrazine imprinted membranes (realized between two glass slides) using a

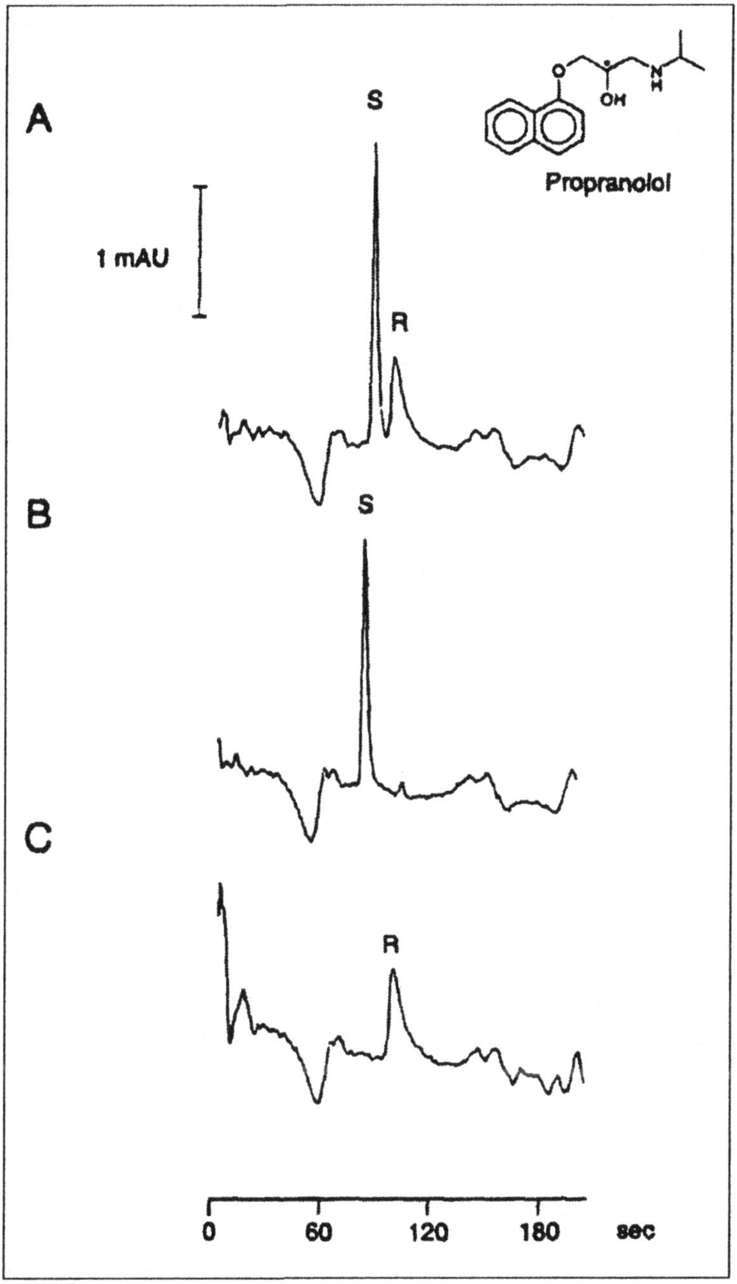

Figure 7.8. Electropherograms of (A) a racemic mixture of propranolol, (B) (S)-propranolol and (C) (R)-propranolol (using in all three experiments the MI capillary prepared using (R)-propranolol as template). Reprinted with permission from: Schweitz L, Andersson LI, Nilsson S. Capillary electrochromatography with predetermined selectivity obtained through molecular imprinting. Anal Chem 1997; 69:1179-1183. © 1997 American Chemical Society

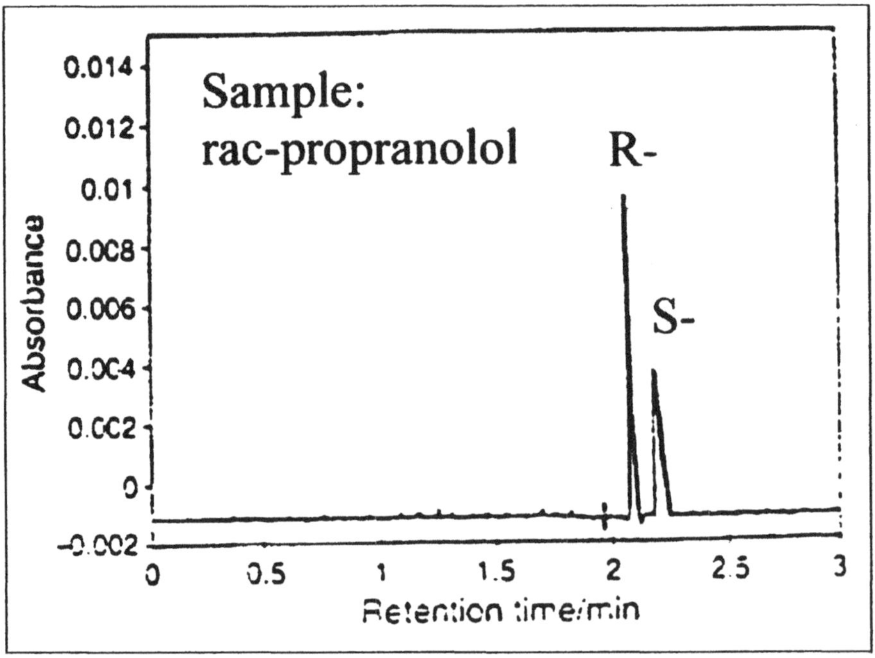

Figure 7.9. Separation of R- and S-propranolol using MIP particles as a chiral additive in the background electrolyte, MIP prepared using S-propranolol as template. Reprinted with permission from: Walshe M, Garcia E, Howarth J et al. Separation of the enantiomers of propranolol by incorporation of molecularly imprinted polymer particles as chiral selectors in capillary electrophoresis. Anal Commun 1997; 34:119-122. © 1997 The Royal Society of Chemistry

reaction mix of MAA/TEGDMA/$CHCl_3$ and oligourethan. Sensors based on these MI membranes were able to detect atrazine conductimetrically within 6-10 min in a concentration below 5 nmol/l.[35,44]

Kobayashi, Wang and coworkers have produced an MI membrane selective for theophylline, using poly(acrylonitrile-coacrylic acid) in DMSO as base polymer and casting the membrane by coagulation in the presence of water.[27,28,29,30] In a first test the membrane did bind the template theophylline (THO) more strongly than the control substance caffeine (CAF), which differs structurally from THO by just one methyl group. A relatively high separation factor $\alpha_{THO/CAF}$ of 52 was obtained, when a copolymer with an optimum content of 15 mol% acrylic acid was used. Both lower and higher acrylic acid contents in the copolymer led to lower separation factors.[28] Furthermore, Wang et al found that the uptake of THO was dependent on the coagulation temperature used during the casting of the MI membrane. For membranes generated at 10° C the uptake for THO was nearly five times higher than for specimen prepared at 40° C.[30]

A MI membrane suitable for chiral separations was recently introduced by Yoshikawa et al They imprinted Boc-L-Trp in THF using an anchor-tetrapeptide (DIDE, DLDE or DVNE) immobilized on a polymer (polystyrene) / copolymer (acrylonitrile-costyrene) by simply allowing the solvent to evaporate to yield the membrane.[13,14,15,16,17] The separation factors obtained with these membranes were between 1.0 and 1.4 and hence not all that impressive. The authors demonstrated that an increase of the Boc-L-Trp/DIDE ratio led to an increase of the adsorbed amounts of both the template (L-Trp) *and* its enantiomer (D-Trp), probably due to an enlargement of the permeation path as a result of the membrane preparation process. At the same time the selectivity decreased.[18]

Mathew-Krotz et al generated film-like membranes imprinted with 9-ethyladenine on glass slides with MAA as functional monomer and EDGMA as cross-linker in DMF as porogen. When evaluating the membranes, they observed an accelerated transport of only adenine or adenine containing compounds through the pores. Analysis of mixtures containing adenosine and guanosine using the 9-ethyladenine imprinted membrane gave a selectivity factor of 3.4 with adenosine passing by much faster than guanosine. The authors explained this result with the low affinity of the membrane for nonadenine-containing compounds together with a high affinity for adenine and its relatives, resulting in a selective transport due to reversible complexations and exchanges between adenine and the binding sites.[38]

Binding Assays Using MIP

MIP have been used for competitive binding assays since 1993, when Vlatakis et al presented a first MIP based radioimmunoassay.[3] In binding assays, as in the analytical bioseparation methods discussed above, MIP are considered as interesting alternatives to true bioligands, such as the antibodies commonly used in immunoassays, mainly due to their superior stability and also to the fact that their generation is comparatively fast and easy. It is hoped that cheap biomimetic MIP could eventually replace the expensive antibodies as receptors. Generally the bottleneck in competitive binding assays is the necessity of using a labeled competitor to the target analyte. For obtaining a competitor in MI-radioimmunoassays the template has to be labeled with a radioisotope which requires special precautions and safety equipment and is furthermore not environmental friendly. In fluorescence based techniques a fluorescent derivative of the analyte has to be prepared without changing the overall molecular structure to much in order to prevent creating pronounced selectivity for either the labeled or the unlabeled molecule in the MIP.

Haupt et al have developed a competitive fluoroimmunoassay based on MIP. They used the herbicide 2,4-dichlorophenoxyacetic acid (2,4-D) (Fig. 7.10, insert) as the template and prepared the MIP in an unusual protic porogen mixture of water and methanol. A nonrelated fluorescent compound, 7-carboxymethoxy-4-methylcoumarin (CMMC) was used as competitor, which can bind to the MIP via the phenoxyacetic acid section (Fig. 7.10, insert).[36] The resulting MIP were operative in both aqueous and organic solvents. For the evaluation of the polymer, the MIP were first suspended in a solution of CMMC to allow the competitor to saturate the imprints. Afterwards the charged MIP were incubated in solutions containing 2,4-dichlorophenoxyacetic acid, 4-chlorophenoxyacetic acid (CPOAc) and phenoxyacetic acid (POAc) respectively. After a period of 2 h—sufficient for the competition of the unlabeled and labeled compounds for the binding sites to have reached equilibrium—the concentration of the displaced CMMC was measured in the supernatant. Figure 7.10 shows typical calibration curves for the competition of CPOAc and POAc respectively with CMMC for the imprints in a) aqueous solution (phosphate buffer) and b) an organic solvent (ACN). CPOAc could be detected in a range from 100 nM to 50 μM in phosphate buffer and from 100 nM to 10 μM in ACN. The results have been compared with those of a corresponding radioimmunoassay and showed similar sensitivity and limit of detection. Compared with a conventional binding assay based on chemiluminescence the MI fluorescence format was more sensitive.[45]

The amounts of template available often determine the maximum quantity of MIP obtainable due to the necessity of having relatively high concentrations of template in the imprinting mixture. Ratios of template : functional monomer : cross-linker around 1:4:20 lead to high costs or to the unfeasibility of an approach, if the template is too expensive or just available in small quantities. However, Yilmaz et al demonstrated that polymers can be imprinted

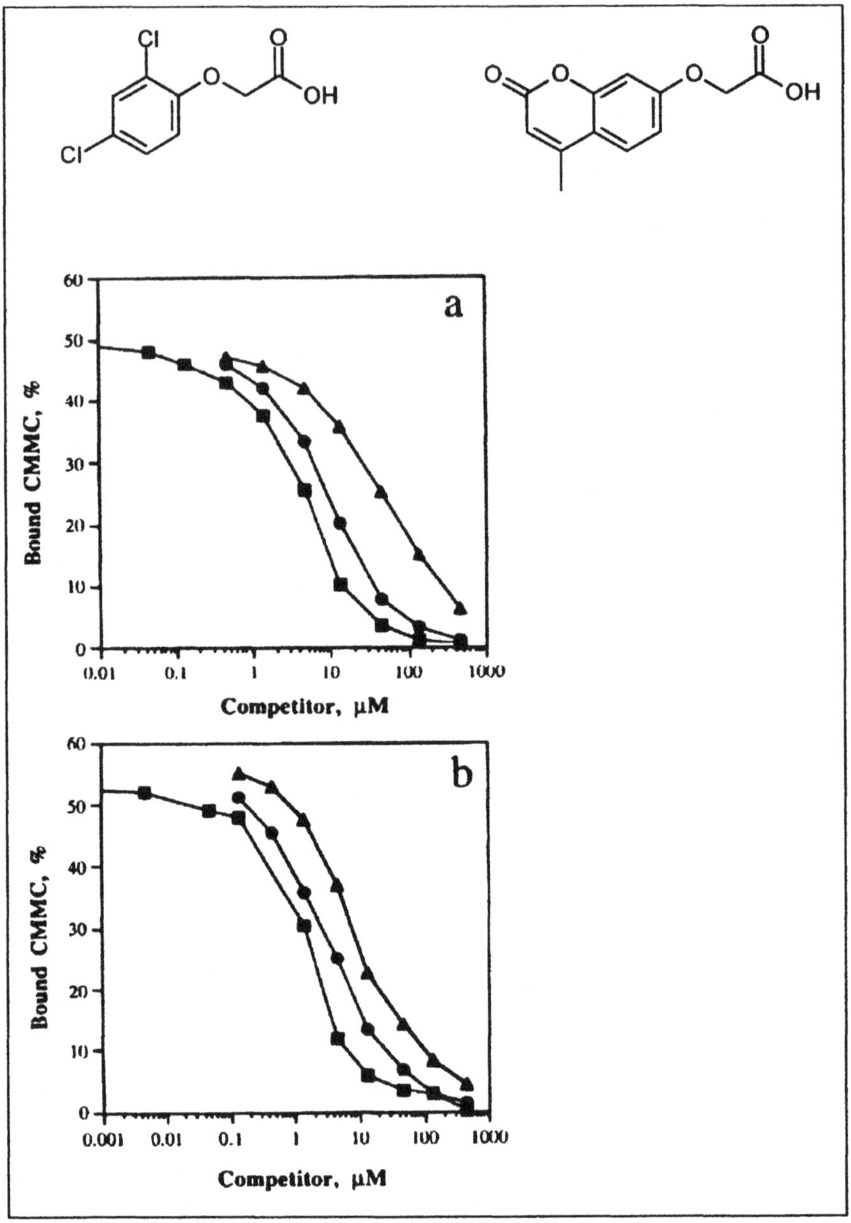

Figure 7.10. Insert: left 2,4-dichlorophenoxyacetic acid (2,4-D), right 7-carboxymethoxy-4-methylcoumarin (CMMC).

Figure 7.10. Standard curves for 2,4-dichlorophenoxyacetic acid (2,4-D). Displacement of CMMC from the 2,4-D-imprinted MIP by 2,4-D (■), CPOAc (●) and POAc (▲) in a) 20 mM phosphate buffer, pH 7, 0.1% Triton X-100 b) acetonitrile. Reprinted with permission from: Haupt K, Mayes AG, Mosbach K. Herbicide assay using an imprinted polymer-based system analogous to competitive fluoroimmunoassays. Anal Chem 1998; 70:3936-3939. © 1998 American Chemical Society

with very small amounts of template, achieving MIP of qualities similar to that of MIP obtained by more conventional approaches. Deviating from the commonly used template : functional monomer ratios of 1:4 or 1:10, they investigated in addition polymers imprinted with theophylline with ratios of 1:500 or even 1:5000. The result was nevertheless a typical MIP with a significantly higher binding capacity for the template than the control polymer. All MIP prepared by the authors showed dissociation constants in the same order of magnitude, i.e., around 10 nM. For the 1:500 MIP a dissociation constant of 14.7 nM was observed and interpreted by the assumption that the excess of functional monomer over template molecules had in fact shifted the equilibrium towards optimal complexation of template during MIP formation (imprinting). The authors also demonstrated that MIP, to be used in binding assays, can be fabricated with unusual low amounts of cross-linkers. MIP in which only 19% of the polymerizable groups were provided by the cross-linker could in regard to the recognition properties compete easily with MIP produced with a more usual degree of cross-linking of 69%.[26]

Ansell et al have combined MI and magnetic bead technology when manufacturing MI polymer beads for drug radioligand binding assays. In this case, the suspension polymerization was carried out in the presence of magnetic iron oxide (magnetite, Fe_3O_4) in a perfluorocarbon. (S)-propranolol was used as template. The selectivity of the imprints was not affected by the magnetism of the material and the beads themselves showed per se no tendency to aggregate because of the fairly low concentration of magnetite particles within the beads. However, once charged with the selectively bound target molecule, the beads could easily be separated from the rest of the mixture by applying a magnetic field.[25]

Sensor Technology

Molecularly imprinted polymers have also been applied for "bio"-sensor development. In this case, the MIP act as specific recognition element linked to a transducer, which generates a signal in response to a binding event taking place at the MIP recognition site.[46] Table 7.5 lists some examples of MIP-based sensors using a variety of transducer types.

Kriz et al developed an optical sensor using a polymer imprinted with dansyl-L-phenylalanine immobilized at the tip of a glass fiber bundle.[51] The authors showed that selective binding of the print molecule induced an corresponding increase in the fluorescence signal. The MIP was generated by using MAA and 4-vinylpyridine as functional monomers to interact with both the amino and the carboxyl functions of the template (dansyl-L-phenylalanine). As shown in Figure 7.11 this led to the formation of chiral cavities with three interaction points interacting preferably with the L-enantiomer of dansyl phenylalanine. However, the time required for reaching equilibrium was 4 hours. Although this is quite long, the simplicity of generating robust MI sensor materials for an unlimited number of different templates constitutes the great advantage of this technique.

MIP to Assist Chemical Synthesis

Catalytic MIP

The idea of generating catalytically active MIP was born, as a result of the production of the first (protein-based) catalytic antibodies (abzymes). The biological antibodies were raised by immunizing mice with so-called transition state analogues (TSA) of specific reactions. The resulting antibody binds with preference to the TSA, thereby stabilizing this particular molecular structure and thus decreasing the activation energy of the corresponding reaction. The major problem in the application of abzymes as catalysts was again the lack in stability of these

Table 7.5. Examples of sensors equipped with MIP as recognition elements

Analyte	Range Examined [μg/ml]	Transducer	Reference
Vitamin K_1	0—4	Ellipsometry	47
Phenylalanine anilide	Qualitative	Capacitance	48
	33—3300	Potentiometry	49
Morphine	0 - 10	Amperometry	50
Dansyl-L-phenylalanine	0—50	Fiber-optic fluorescence	51
Atrazine	0 - 0.5	Conductometry	33,35,44
Benzyltriphenyl-phosphonium ions	0 - 400	Conductometry	52
Sialic acid	0 - 3	Optical fluorescence	53

biological entities. Several groups have since tried to combine the idea of abzymes with the principle of MI. TSA of well-understood catalytic reactions were selected as templates and mixed into the usual MI-cocktail of functional monomer, cross-linker, initiator and porogen, followed by polymerization. The subsequent treatment of the catalytic MIP was similar to that used for generating MIP stationary phases for chromatographic applications.

Leonhardt et al were able to show a specific hydrolytic effect when treating nitrophenyl esters with an MIP imprinted with pyridine derivatives of N-boc-amino acids using DVB and 4(5)-vinylimidazole as monomers in combination with chelated Co^{2+} ions.[54] Compared to the control polymer, hydrolysis was accelerated by a factor of 4 to 5, while a comparison with MIP imprinted with other pyridine derivatives of N-boc-amino acids only gave an acceleration of the reaction by a factor of 2 to 3. Robinson et al used phosphonates as TSA and showed a catalytic effect of an MIP imprinted with p-nitrophenylmethyl phosphonate, using 4(5)-vinylimidazole and Co^{2+} ions, on the hydrolysis of p-nitrophenol acetate. The authors—aware of the fact that imidazole containing polymers in general exhibit catalytic effects—could nevertheless demonstrate that the imprinted specimens were of 60% higher activity than the control polymers.[55]

Phosphonates have also been selected as TSA templates by Okhubo et al,[56,57,58] for the purpose of hydrolyzing amino acid esters. Stereoselective MIP, some of them water soluble, were produced and an investigation of their catalytic properties showed an increase in the reaction rate constants by a factor of more than 3 in the case of pseudo first order rate constants obtained with and without the MIP.[59,60] A highly selective MIP imprinted with (4-carboxybenzyl)-phosphonic acid monoester has been presented by Wulff and coworkers.[61] Using N,N′-diethyl(4-vinylphenyl)amidine as functional monomer, a 100-fold acceleration of the esterolysis of a (4-carboxybenzyl)-carboxyl ester was achieved. No comparison with a (nonimprinted, but otherwise identical) control polymer was attempted. Instead the reaction rates in the presence of the MIP were compared with hydrolysis rates obtained in free solution or in a solution containing (4-ethylphenyl)amidine or in the presence of polymers imprinted with the benzoate of N,N′-diethyl(4-vinylphenyl)amidine).

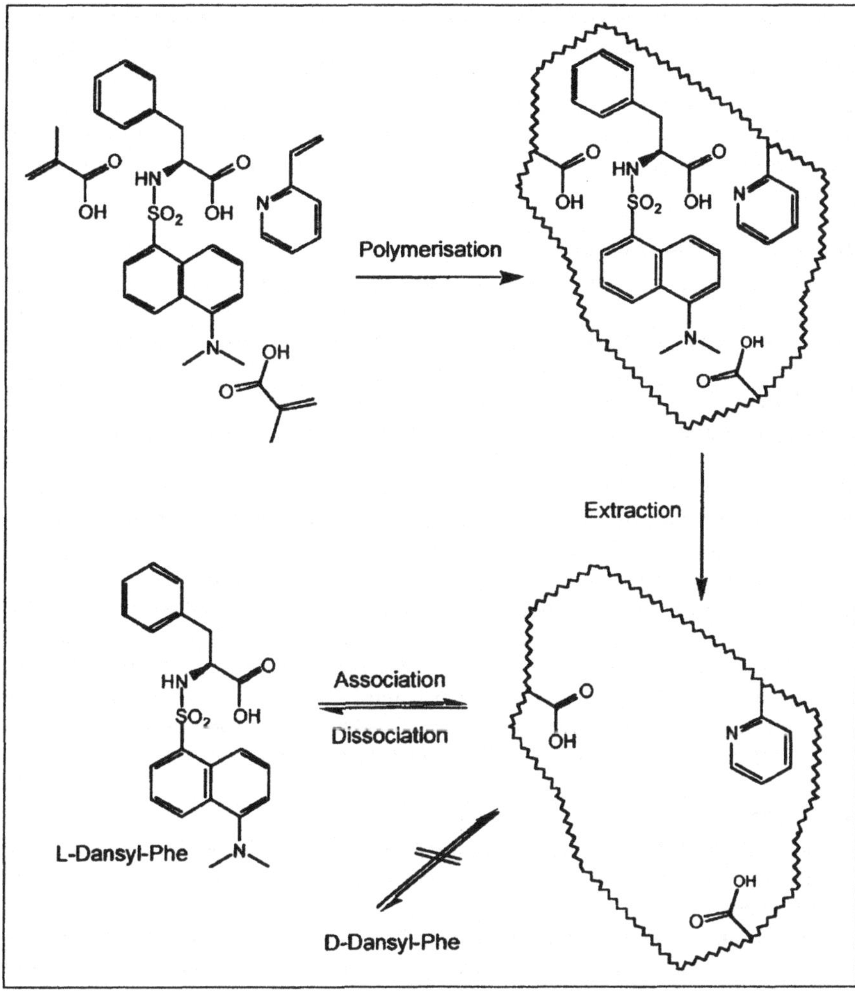

Figure 7.11. Chiral MIP imprinted with dansyl-L-phenylalanine showing selectivity towards the L-enantiomer of dansyl phenylalanine. Reprinted with permission from: Kriz D, Ramström O, Mosbach K. Molecular imprinting—New possibilities for sensor technology. Anal Chem 1997; 345A-349A. © 1997 American Chemical Society

Sellergren et al have developed enantioselective MIP-catalysts using a phosphonate as template. For the determination of the catalytic properties they compared the MIP and a control polymer. Rate constant ratios (k_{MIP}/k_{CP}) of up to 2.54 were observed.[62]

Catalytic MIP for the dehydrofluorination of 4-fluoro-4-(p-nitrophenyl)-2-butanone have been generated using different kind of templates including N-benzyl-isopropylamine,[63] benzylmalonic acid,[64] N-methyl-N-(4-nitrobenzyl)-δ-aminovaleric acid[65] or N-isopropyl-4-nitrobenzylamine.[66] Imprinting in aqueous solution was used in some of these approaches.[65,66] Polymers imprinted with chlorendic anhydride (MAA and EGDMA as monomers) could be used as catalysts for the Diels-Alder reaction of tetrachlorothiophene dioxide (TCTD) with maleic anhydride. Acceleration factors ($(k_{imp}—k_{nonimp}) / k_{uncat}$) of up to 270 have been observed when the MIP was compared with a control polymer and a polymer-free solution.[67]

Santora et al developed an MIP, which was capable of accelerating a similar reaction, i.e., linking cyclohexadiene with a α β-unsaturated carbonyl compounds (cyclic Diels-Alder reaction).[68] Furthermore, redox reactions,[69] transacylations[70] or the synthesis of lactons[71] have been catalyzed by MIP (see also Table 7.6). For the aldol-condensation of acetophenone with benzaldehyde, a polymer imprinted with a dibenzoylmethane/Co^{2+} complex was utilized.[72] These approaches may just be the beginning of a broad implementation of catalytically active MIP in chemical synthesis.

MIP for Continuous Product Removal

Continuous removal of the product has been proposed as a means to "shift" the equilibrium of reversible reactions towards the desired side. Once more MIP can be seen as a generic solution to the problem of finding a highly specific yet robust adsorber for the target substance. A few examples of using MIP successfully for that purpose can be found in the literature. Ye et al were, for example, among the few to apply a product-imprinted polymers for shifting the reaction equilibrium of a synthesis towards product formation by removing the product continuously from the solution.[73] As a model system, the authors chose an enzymatic conversion of Z-L-Asp with L-Phe-OMe to Z-α-aspartame, a precursor of the sweetener α-aspartame, Figure 7.12, insert. It was demonstrated that a MIP imprinted with the product Z-α-aspartame adsorbs the product during the reaction, thus permanently ensuring a low level of Z-α-aspartame in the reaction mixture and consequently increasing the overall yield. Figure 7.12 shows the effects of different additives. The highest yield of product was obtained when using the MIP imprinted with the product. Nevertheless, even utilization of the control polymer (no template used) caused an increase of the product yield (explained by nonspecific adsorption of the product) as well as the MIP imprinted with Z-L-Asp and L-Phe-OMe (1:1), i.e., the substrates of the reaction. However, after an incubation time of 48 h the effect of nonspecific binding became less important, due to the release of the nonspecifically bound product.

Conclusions

Admittedly, today's MIP often do not perform yet at par with established materials like, for example, β-cyclodextrins for enantioselective separations. However, they did improve considerably over the last 5-10 years and already constitute useful alternatives in some areas, for example as adsorbents in solid phase extraction. As such they can be applied for sample preparation or preconcentration in case of complex matrices or a high number of other contaminating substances.

Compared to true biological materials the biomimetic MIP have a number of advantages:

1. Cost of Production: Due to the use of inexpensive monomers, porogens etc. the MIP itself is relatively cheap; a costly template can be a limiting factor.
2. Stability: MIP show good resistance against both physical and chemical factors, including mechanical stress, high pressures and aggressive/organic solvents.
3. Durability: MIP can be reused several hundred times; they can be stored under dry conditions without any noticeable loss of performance.
4. Flexibility/versatility: MIP can be imprinted with an almost unlimited number of different templates; only the requirement of distinctive functional groups on the template can be a limiting factor.

Current disadvantages of MIP include:

1. Heterogeneity of the imprints: The imprints obtained in the polymer network show varied specificity; this causes, for example, broad peaks in chromatographic separations.
2. The "H_2O-problem": Water is often a strong competitor for the interaction between the functional monomers and the template and therefore has to be excluded from the imprinting mixture. However, many templates (peptides/proteins) in native form are soluble only in water.

Table 7.6. Important reactions catalyzed by MIP

TSA	Substrate	Relative Catalytic Effect of the MIP	Reference
		Hydrolysis	
Pyridin-derivatives of N-Boc-amino acids	Nitrophenyl ester	$k_{imp} / k_{nonimp} = 4\text{-}5$	54
p-Nitrophenylmethyl phosphonate	p-Nitrophenol acetate	$k_{imp} / k_{nonimp} = 1.6$	55
Phosphonate	Amino acid ester	$k_{imp} / k_{uncat} = 3$	59,60
Phosphonate	Carbonic acid ester	$k_{imp} / k_{uncat} \sim 100$	61
Phosphonate	Amino acid ester	$k_{imp} / k_{nonimp} = 2.54$	62
		Dehydrofluorination	
N-Benzyl-isopropylamine	4-Fluoro-4-(p-nitro-phenyl)-2-butanone	$k_{imp} / k_{nonimp} < 2.4$	63
Benzylmalonic acid	4-Fluoro-4-(p-nitro-phenyl)-2-butanone	$k_{imp} / k_{control\text{-}imp} < 3.5$ (12.8)	64
N-methyl-N-(4-nitrobenzyl)-δ-aminovaleric acid	4-Fluoro-4-(p-nitro-phenyl)-2-butanone	$k_{imp} / k_{uncat} = 3.3$	65
		Diels-Alder Reaction	
Chlorendic anhydride	Tetrachlorothiophen-dioxide + maleic anhydride	$(k_{imp} - k_{nonimp}) / k_{uncat} = 270$	67
		Aldol Condensation	
Dibenzoylmethane (DBM) + Co^{2+}	Acetophenone and benzaldehyde	$r_{(DBM/Co^{2+}\text{-}MIP)} / r_{(Co^{2+}\text{-}MIP)}$ £ 2	72

Rate constants

	Determined in the presence of:
k_{imp}	the imprinted polymer
k_{nonimp}	the nonimprinted polymer
$k_{control\text{-}imp}$	MIP, imprinted with a template similar to the original one
k_{uncat}	no catalyst used

Rates

$r_{(DBM/Co^{2+}\text{-}MIP)}$	MIP, imprinted with template and Co^{2+}
$r_{(Co^{2+}\text{-}MIP)}$	MIP, imprinted with only Co^{2+}

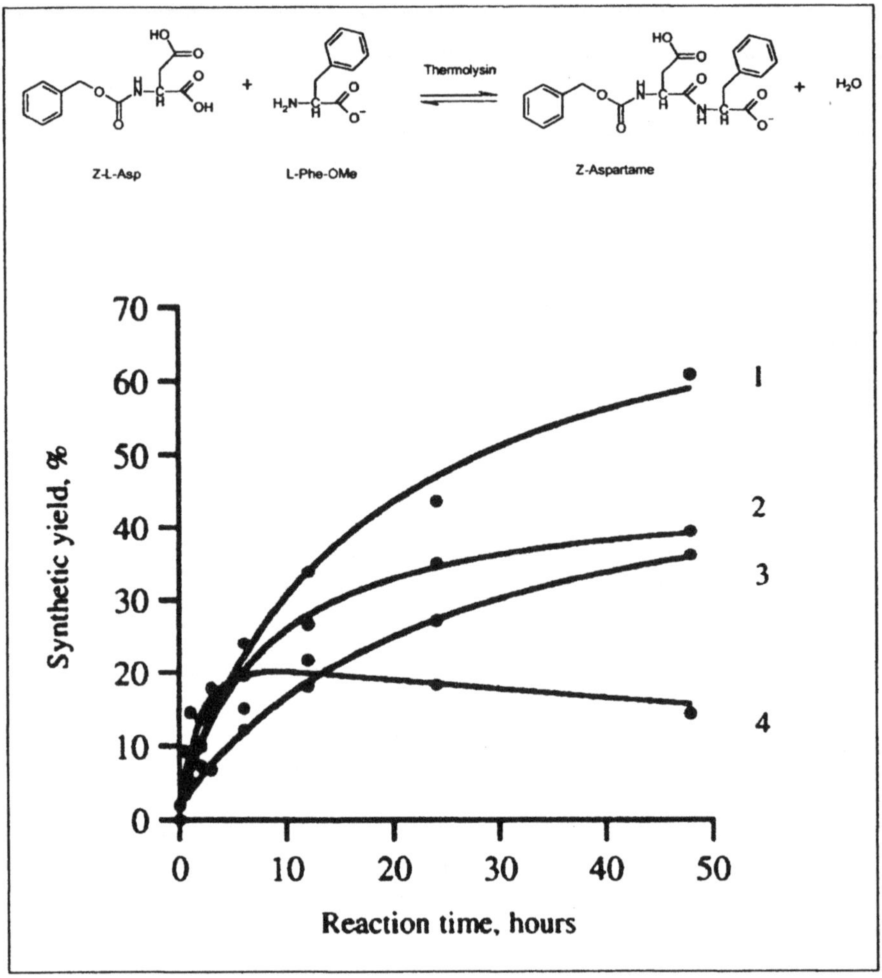

Figure 7.12. Synthesis of Z-aspartame in the presence and absence of MIP imprinted with product: curve 1: MIP_1, Z-aspartame as print molecule, curve 2: control polymer, no print molecule, curve 3: MIP_2, Z-L-Asp and L-Phe-OMe as print molecules, curve 4: the polymer-free enzymatic reaction. Reprinted with permission from: Ye L, Månsson M-O, Mosbach K et al. A new application of molecularly imprinted materials. J Molec Recog 1998; 11:75-78. © 1998 John Wiley & Sons, Ltd.

3. Efficiency: Many MIP are prone to nonspecific interactions caused by randomly arranged interactive groups on the polymer surface (surplus of functional monomer) which reduces the selectivity.
4. Prolonged bleeding of template: It has been observed that sometimes a 100% extraction of the template from the MIP after imprinting is not possible. However, the remaining template tends to leak out with time during the subsequent use of the MIP.[74]

However, it should be noted that especially in the area of binding assays using labeled molecules, MIP-based assays already perform with similar efficiency as conventional immunoassays. A very recent application of MIP is their use to screen combinatorial libraries, e.g., for interactive molecules. Ramström et al used MIP imprinted with androsterone structures for selecting related compounds from a combinatorial steroid library. The molecules used as templates could be selectively fished out from mixtures of 12 different steroids, demonstrating the usefulness of highly selective MIP as biomimetic receptors.[75]

The imprinting of large molecules like proteins,[5,76] DNA and even of cells[77] has just begun and currently the predominant number of publications in the area of MIP is about using small molecules as templates. However, Shi et al recently published about imprinting surfaces with protein-recognition sites.[76] In a first step, proteins like albumin and lysozyme were coated with disaccharides and then used to imprint a polymeric film formed by a technique called radio-frequency glow-discharge plasma deposition. After removal of the proteins, the remaining cavities in the polymer film still contained the carbohydrate layer, which rendered these sites capable of specifically recognizing the print protein in a competitive assay. It was indicated that such a sugar coat enables multivalent interactions with the template due to the increased number of possible hydrogen bonds, thus increasing the affinity of the imprinted phase for the template. This work represents a novel and important approach towards molecular imprinting of large biomolecules.

The fact that more and more groups worldwide are focusing on MIP in general is demonstrated by the exponentially growing number of publications in that field. It has to be shown in the near future that MIP are real competitors for existing "natural" biospecific materials, in analytical chemistry, chemical synthesis or catalysis. For the generation of more reliable and powerful MIP the principle of the biological interactions, which serve as model have to be understood in even more detail to enable efficient mimicking. If these hurdles can be taken, MIP stand an excellent chance of becoming important industrial product for many areas, including medicine, chemistry and biotechnology.

Abbreviations

ABCHC	Azobis(cyclohexanecarbonitrile)
ABDV	2,2′-Azobis(2,4-dimethyl valeronitrile)
ACN	Acetonitrile
AIBN	Azobis(isobutyronitrile)
CE	Capillary electrophoresis
CEC	Capillary electrochromatography
CP	Control polymer
CPOAc	4-Chlorophenoxyacetic Acid
DMF	Dimethylformamide
DMSO	Dimethylsulfoxide
DVB	Divinylbenzene
EGDMA	Ethylenglycol dimethacrylate
HPCE	High performance capillary electrophoresis
HPLC	High performance liquid chromatography
MAA	Methacrylic acid
MI	Molecular imprinting / Molecularly imprinted
MIP	Molecular pmprinted polymer
PFPS	Perfluoro polymeric surfactant
POAc	Phenoxyacetic acid
TCTD	Tetrachlorothiophene dioxide
TEGDMA	Triethylenglycol dimethacrylate
THF	Tetrahydrofuran
THO	Theophylline
TRIM	Trimethylolpropane trimethacrylate
TSA	Transition state analogue

References

1. Pauling L. A theory of the structure and process of formation of antibodies. J Am Chem Soc 1940; 62:2643-2657.
2. Wulff G, Sarhan A. Angew. Chem Int Ed Engl 1972; 11:341-341.
3. Vlatakis G, Andersson LI, Müller R et al. Drug assay using antibody mimics made by molecular imprinting. Nature 1993; 361:645-647.
4. Wulff G. Molecular imprinting in cross-linked materials with the aid of molecular templates—A way towards artificial antibodies. Angew Chem Int Ed Engl 1995; 34:1812-1832.
5. Kempe M, Mosbach K. Separation of amino acids, peptides and proteins on molecularly imprinted stationary phases. J Chromatogr 1995; 691:317-323.
6. Mosbach K, Ramström O. The emerging technique of molecular imprinting and its future impact on biotechnolgy. Bio/Technol 1996; 14:163-170.
7. Vidyasankar S, Ru M, Arnold FH. Molecularly imprinted ligand-exchange adsorbents for the chiral separation of underivatized amino acids. J Chromatogr 1997; 775(1-2):51-63.
8. Wulff G, Vesper W. Preparation of chromatographic sorbents with chiral cavities for racemic resolution. J Chromatogr 1978; 167:171-186.
9. Kempe M, Mosbach K. Receptor binding mimetics: A novel molecularly imprinted polymer. Tetrahedron Letters 1995; 36:3563-3566.
10. Kempe M. Antibody-mimicking polymers as chiral stationary phases in HPLC. Anal Chem 1996, 68:1948-1953.
11. Yu C, Mosbach K. Molecular imprinting utilizing an amide functional group for hydrogen bonding leading to highly efficient polymers. J Org Chem 1997; 62(12):4057-4064.
12. Lin JM, Nakagama T, Uchiyama K et al. Molecularly imprinted polymer as chiral selector for enantioseparation of amino acids by capillary gel electrophoresis. Chromatographia 1996; 43(11-12):585-591.
13. Yoshikawa M, Izumi J, Kitao T et al. Molecularly imprinted polymeric membranes for optical resolution. J Membr Sci 1995; 108:171-175.
14. Yoshikawa M, Fujisawa T, Izumi J et al. Chiral recognition of N-α-acetyltryptophans with molecularly imprinted polymeric membranes containing DVNE derivatives. Sen-I Gakkaishi 1998; 54(2):77-84.
15. Yoshikawa M, Fujisawa T, Izumi J et al. Molecularly imprinted polymeric membranes involving tetrapeptide EQKL derivatives as chiral-recognition sites toward amino acids. Anal Chim Acta 1998; 365:59-67.
16. Yoshikawa M, Izumi J, Ooi T et al. Carboxylated polysulfone membranes having a chiral recognition site induced by an alternative molecular imprinting technique. Polymer Bull 1998; 40:517-524.
17. Yoshikawa M, Ooi T, Izumi J. Alternative molecularly imprinted membranes from a derivative of natural polymer, cellulose acetate. J Appl Polym Sci 1999; 72:493-499.
18. Yoshikawa M, Izumi J, Kitao T et al. Molecularly imprinted polymeric membranes containing DIDE derivatives for optical resolution of amino acids. Macromolecules 1996; 29:8197-8203.
19. Mayes AG, Mosbach K. Molecularly imprinted polymer beads: Suspension polymerization using a liquid perfluorocarbon as the dispersing phase. Anal Chem 1996; 68:3769-3774.
20. Kempe M, Glad M, Mosbach K. An approach towards surface imprinting using the enzyme ribonuclease A. J Mol Recogn 1995; 8:35-39.
21. Skudar K, Brüggemann O, Wittelsberger A et al. Selective recognition and separation of β-lactam antibiotics using molecularly imprinted polymers. Anal Commun 1999; 36:327-331.
22. Schweitz L, Andersson LI, Nilsson S. Capillary electrochromatography with predetermined selectivity obtained through molecular imprinting. Anal Chem 1997; 69:1179-1183.
23. Nilsson S, Schweitz L, Andersson LI, Three approaches to enantiomer separation of β-adrenergic antagonists by capillary electrochromatography. Electrophoresis 1997; 18:884-890.
24. Walshe M, Garcia E, Howarth J et al. Separation of the enantiomers of propranolol by incorporation of molecularly imprinted polymer particles as chiral selectors in capillary electrophoresis. Anal Commun 1997; 34:119-122.
25. Ansell RJ, Mosbach K. Magnetic molecularly imprinted polymer beads for drug radioligand binding assay. Analyst 1998; 123:1611-1616.

26. Yilmaz E, Mosbach K, Haupt K. Influence of functional and cross-linking monomers and the amount of template on the performance of molecularly imprinted polymers in binding assays. Anal Commun 1999; 36:167-170.
27. Kobayashi T, Wang HY, Fujii N. Molecular imprinting of theophylline in acrylonitrile-acrylic acid copolymer membrane. Chem Lett 1995:927-928.
28. Kobayashi T, Wang HY, Fujii N. Molecular imprint membranes of polyacrylonitrile copolymers with different acrylic acid segments. Anal Chim Acta 1998; 365:81-88.
29. Wang HY, Kobayashi T, Fujii N. Surface molecular imprinting on photosensitive dithiocarbamoyl polyacrylonitrile membranes using photograft polymerization. J Chem Tech Biotechnol 1997; 70:355-362.
30. Wang HY, Kobayashi T, Fukaya T et al. Molecular imprint membranes prepared by the phase inversion precipitation technique. 2. Influence of coagulation temperature in the phase inversion process on the encoding in polymeric membranes. Langmuir 1997; 13:5396-5400.
31. Sellergren B. Imprinted dispersion polymers: A new class of easily accessible affinity stationary phases. J Chromatogr 1994; 673:133-141.
32. Brüggemann O, Freitag R, Whitcombe MJ et al. Comparison of polymer coatings of capillaries for capillary electrophoresis with respect to their applicability to molecular imprinting and electrochromatography. J Chromatogr 1997; 781:43-53.
33. Piletsky SA, Piletskaya EV, Elgersma AV et al. Atrazine sensing by molecularly imprinted membranes. Biosens Bioelectron 1995; 10:959-964.
34. Piletsky SA, Piletskaya EV, Panasyuk TL et al. Imprinted membranes for sensor technology: Opposite behavior of covalently and noncovalently imprinted membranes. Macromolecules 1998; 31:2137-2140.
35. Sergeyeva TA, Piletsky SA, Brovko, et al. Conductimetric sensor for atrazine detection based on molecularly imprinted polymer membranes. Analyst 1999; 124:331-334.
36. Haupt K, Mayes AG, Mosbach K. Herbicide assay using an imprinted polymer-based system analogous to competitive fluoroimmunoassays. Anal Chem 1998; 70:3936-3939.
37. Norrlöw O, Månsson M-O, Mosbach K. Improved chromatography: Prearranged distances between boronate groups by the molecular imprinting approach. J Chromatogr 1987; 396:374-377.
38. Mathew-Krotz J, Shea KJ. Imprinted polymer membranes for the selective transport of targeted neutral molecules. J Am Chem Soc 1996; 118:8154-8155.
39. Nilsson KGI, Sakaguchi K, Gemeiner P et al. Molecular imprinting of acetylated carbohydrate derivatives into methacrylic polymers. J Chromatogr 1995; 707:199-203.
40. Wulff G, Haarer J. Enzyme-analogue built polymers. The preparation of defined chiral cavities for the racemic resolution of free sugars. Makromol Chem.1991; 192:1329-1338.
41. Whitcombe MJ, Rodriguez ME, Villar P et al. A new method for the introduction of recognition site functionality into polymers prepared by molecular imprinting: Synthesis and characterization of polymeric receptors for cholesterol. J Am Chem Soc 1995; 117:7105-7111.
42. Norrlöw O, Glad M, Mosbach K. Acrylic polymer preparations containing recognition sites obtained by imprinting with substrates. J Chromatogr 1984; 299:29-41.
43. Schweitz L, Andersson LI, Nilsson S. Molecular imprint-based stationary phases for capillary electrochromatography. J Chromatogr 1998; 817:5-13.
44. Sergeyeva TA, Piletsky SA, Brovko, et al. Selective recognition of atrazine by molecularly imprinted polymer membranes. Development of conductometric sensor for herbicides detection. Anal Chim Acta 1999; 392:105-111.
45. Haupt K. Molecularly imprinted sorbent assays and the use of nonrelated probes. React Funct Polym 1999; 41:125-131.
46. Kriz D, Ramström O, Mosbach K. Molecular imprinting—New possibilities for sensor technology. Anal Chem 1997:345A-349A.
47. Andersson L, Mandenius C, Mosbach K. Studies on guest selective molecular recognition on an octadecyl silylated silicon surface using ellipsometry. Tetrahedr Lett 1988; 29:5437-5440.
48. Hedborg E, Winquist F, Lundström I et al. Some studies of molecularly-imprinted polymer membranes in combination with field-effect devices. Sens Actuators A 1993; 37:796-799.

49. Andersson LI, Miyabayashi A, O'Shannessy DJ et al. Enantiomeric resolution of amino acid derivatives on molecularly imprinted polymers as monitored by potentiometric measurements. J Chromatogr 1990; 516:323-331.
50. Kriz D, Mosbach K. Competitive amperometric morphine sensor based on an agarose immobilized molecularly imprinted polymer. Anal Chim Acta 1995; 300:71-75.
51. Kriz D, Ramström O, Svensson A et al. Introducing biomimetic sensors based on molecularly imprinted polymers as recognition elements. Anal Chem 1995; 67:2142-2144.
52. Kriz D, Kempe M, Mosbach K. Introduction of molecularly imprinted polymers as recognition elements in conductometric chemical sensors. Sens Actuators B 1996; 33:178-181.
53. Piletsky SA et al. Anal Lett 1996; 29:157-170.
54. Leonhardt A, Mosbach K. Enzyme-mimicking polymers exhibiting specific substrate binding and catalytic functions. React Polym 1987; 6:285-290.
55. Robinson DK, Mosbach K. Molecular imprinting of a transition state analogue leads to a polymer exhibiting esterolytic activity. J Chem Soc Chem Commun 1989:969-970.
56. Ohkubo K, Urata Y, Hirota S et al. Homogeneous esterolytic catalysis of a polymer prepared by molecular imprinting of a transition state analogue. J Mol Catal 1994; 93:189-193
57. Ohkubo K, Urata Y, Honda Y et al. Preparation and catalytic property of L-histidyl group-introduced, crosslinked poly(ethylene imine)s imprinted by a transition-state analogue of an esterolysis reaction. Polymer 1994; 35(24):5372-5374.
58. Ohkubo K, Urata Y, Hirota S et al. Homogeneous and heterogeneous esterolytic catalyses of imidazole-containing polymers prepared by molecular imprinting of a transition state analogue. J Mol Catal 1994; 87:L21-L24.
59. Ohkubo K, Urata Y, Hirota S et al. Catalytic activities of novel L-histidyl group-introduced polymers imprinted by a transtition state analogue in the hydrolysis of amino acid esters. J Mol Catal 1995; 101:L111-L114.
60. Ohkubo K, Funakoshi Y, Sagawa T. Catalytic activity of a novel water-soluble cross-linked polymer imprinted by a transition-state analogue for the stereoselective hydrolysis of enantiomeric amino acid esters. Polymer 1996; 37(17):3993-3995.
61. Wulff G, Gross T, Schönfeld R. Enzyme models based on molecularly imprinted polymers with strong esterase activity. Angew Chem Int Ed Engl 1997; 36:1961-1964.
62. Sellergren B, Shea KJ. Enantioselective ester hydrolysis catalyzed by imprinted polymers. Tetrahedron 1994; 5(8):1403-1406.
63. Müller R, Andersson LI, Mosbach K. Molecularly imprinted polymers facilitating a β-elimination reaction. Makromol Chem Rapid Commun 1993; 14:637-641.
64. Beach JV, Shea KJ. Designed catalysts. A synthetic network polymer that catalyzes the dehydrofluorination of 4-fluoro-4-(p-nitrophenyl)butan-2-one. J Am Chem Soc 1994; 116:379-380.
65. Ohya Y, Miyaoka J, Ouchi T. Recruitment of enzyme activity in albumin by molecular imprinting. Macromol Rapid Commun 1996; 17:871-874.
66. Slade CJ, Vulfson EN. Induction of catalytic activity in proteins by lyophilization in the presence of a transition state analogue. Biotech Bioeng 1997; 57(2):211-215.
67. Liu X-C, Mosbach K. Studies towards a tailor-made catalyst for the Diels-Alder reaction using the technique of molecular imprinting. Macromol Rapid Commun 1997; 18:609-615.
68. Santora BP, Larsen AO, Gagne MR. Toward the molecular imprinting of titanium Lewis acids: demonstration of Diels-Alder catalysis. Organometallics 1998; 17:3138-3140.
69. Locatelli F, Gamez P, Lemaire M. Molecular imprinting of polymerized catalytic complexes in asymmetric catalysis. J Mol Catal 1998; 135:89-98.
70. Matsuishi T, Shimada T, Morihara K. Definitive evidence for enantioselective catalysis over "molecular footprint" catalytic cavities chirally imprinted on a silicat(alumina) gel surface. Chem Lett 1992:1921-1924.
71. Heilmann J, Maier WF. Problems in selective catalysis with molecular imprints in silica—Selective lactones formation from hydroxyesters in micropores-.Z. Naturforsch 1995; 50b:460-468.
72. Matsui J, Nicholls IA, Karube I et al. Carbon-carbon bond formation using substrate selective catalytic polymers prepared by molecular imprinting: An artificial class II aldolase. J Org Chem 1996; 61:5414-5417.

73. Ye L, Ramström O, Månsson M-O et al. A new application of molecularly imprinted materials. J Molec Recog 1998; 11:75-78.
74. Andersson LI, Papricia A, Arvidsson T. A highly selective solid phase extraction sorbent for preconcentration of sameridine made by molecular imprinting. Chromatographia 1997; 46:57-62.
75. Ramström O, Ye L, Krook M et al. Screening of a combinatorial steroid library using molecularly imprinted polymers. Anal Commun 1998; 35:9-11.
76. Shi H, Tsai W-B, Garrison MD et al. Template-imprinted nanostructured surfaces for protein recognition. Nature 1999; 398:593-597.
77. Aherne A, Alexander C, Payne MJ et al. Bacteria-mediated lithography of polymer surfaces. J Am Chem Soc 1996; 118:8771-8772.

Index

S

T

For Product Safety Concerns and Information please contact our EU representative GPSR@taylorandfrancis.com Taylor & Francis Verlag GmbH, Kaufingerstraße 24, 80331 München, Germany

Printed and bound by CPI Group (UK) Ltd, Croydon, CR0 4YY
14/09/2026
02216243-0020